The Foreseeable Future for Water Planning

Time to Change

KWR Watercycle Research Institute Series

The Foreseeable Future for Water Planning
Time to Change

Edited by: Andrew Segrave

With contributions by:
Andrew Segrave
Mariëlle van der Zouwen
Wim van Vierssen
Bei Wen

IWA PUBLISHING

Published by

IWA Publishing
Alliance House
12 Caxton Street
London SW1H 0QS, UK
Telephone: +44 (0)20 7654 5500
Fax: +44 (0)20 7654 5555
Email: publications@iwap.co.uk
Web: www.iwapublishing.com

First published 2014

British Library Cataloguing in Publication Data
A CIP catalogue record for this book is available from the British Library

ISBN: 9781780406121 (Hardback)
ISBN: 9781780406138 (eBook)

KWR Watercycle Research Institute

"Bridging science to practice" – that is the mission of KWR Watercycle Research Institute. Our shareholders are the ten Dutch water companies. KWR researchers work at the interface of science, business and society. Their strength lies in their ability to translate scientific knowledge into practical and implementable solutions for end-users in the Dutch and European water sectors. KWR has developed a solid reputation as a top-level innovation accelerator and international network builder.

Contents

Biographical Note

ANDREW SEGRAVE spearheads scientific futures studies at KWR Watercycle Research Institute. He is fascinated by people's views towards the future and how this influences their behaviour. As coordinator responsible for the thematic research on trends and future perspectives for the Joint Water Sector Research Programme of the Dutch water companies, and through his work as a consultant, Andrew also has much experience at applying methods for horizon scanning and strategic planning in practice.

MARIËLLE VAN DER ZOUWEN is head of the knowledge management division at KWR Watercycle Research Institute. Her management and research activities focus on complex knowledge management models and networks in the water sector and knowledge practices as social interfaces between actors from science, society and industry. Previous experience includes research and teaching about multi-actor and multi-level policy processes and governance trends at Wageningen University and Radboud University in the Netherlands.

WIM VAN VIERSSEN is CEO of KWR Watercycle Research Institute, professor in Science System Assessment of water related research at Delft University of Technology. His multi-disciplinary research currently concerns the complex, dynamic interactions between actors from science, society and industry and the rules that govern them.

BEI WEN works for Elsevier's Science and Technology Books in the role of Senior Web Analyst. She is a PhD candidate in Science System Assessment of water related research at Delft University of Technology, as well as a guest scientific researcher at KWR Watercycle Research Institute. Her research focuses on the structure and dynamics of knowledge networks in the water research field to provide insights that are instrumental in facilitating intersectoral and interdisciplinary collaborations. Her areas of expertise include: data mining and knowledge discovery, big data analysis & visualization, the use and development of scientometric indicators, and social network analysis.

Acknowledgements

KWR Watercycle Research Institute provided funding and time for the entire project, in part through projects financed by the Joint Research Programme of the Dutch Water Companies. We would like to extend our sincere gratitude to the people who generously opened the door to local contacts in Brazil, Ghana, Japan, and the Netherlands, to those who assisted in planning the travels and interviews, to the translators, to those who were so hospitable during our visits, and finally to all of the interviewees who shared with us their Time Perspectives.

Summary

The decisions people make, and the actions they take, depend on how they conceptualize and experience time. This fundamental and influential factor is seldom acknowledged, little understood, and rarely considered explicitly in planning; be that for the material systems or the knowledge systems in the water sector. The objectives of the research that is described in this were (1) to characterize similarities and differences in the Time Perspectives of people working in the water sector worldwide and (2) to theorize about how this temporal ambiguity influences, and can be used for, water planning and setting research agendas.

First an interpretive framework was developed including a model of Time Perspective, which was designed to make clear the most fundamental assumptions about the (relationships between) aspects of an individual's Time Perspective that influence the decisions they make and the actions they take. This model builds on explicit foundations from Critical Realism and brings together psychological theories of motivation and internal time consciousness. The model formed a framework that was used to develop the Foreseeable Future Multi-measure Method. This new method was designed to characterize relevant aspects of an individual's Time Perspective in an integrated fashion. It is a structured interview with five successive steps that are sequenced to account for socially acceptable responses and to facilitate triangulation:

1. Describe the goals and events that motivate the interviewee's decisions and actions
2. Characterize the structure and orientation of the interviewee's concept of time
3. Measure the temporal extension of the interviewee's motivational objects
4. Triangulate the indirect measures with direct questions by completion of sentences
5. Triangulate the preceding measures with an inventory for typifying orientation

The Foreseeable Future Multi-measure Method was used to interview 309 managers, practical workers, and scientists in the Netherlands, Ghana, Brazil, and Japan. The author of this book conducted all of the interviews in person together with a local interpreter in each country. The four countries were chosen to represent the international diversity in Time Perspectives and the three professional roles provided a sample of the diversity in the level of abstraction of water professionals from theory to practice.

The most noteworthy conclusions of the empirical study concern differences in the temporal extent of the goals and events that motivate people. Scientists are motivated by objects that lie furthest into the future followed by managers and then practical workers (Group medians: 8.4; 3.5; 1.7 years). Across national cultures, the time horizons of the Japanese and Brazilian interviewees were found to be longest. The differences across professional roles are greater and more significant than those across countries. But the characteristics of national cultures were found to amplify the differences across professional roles. Within the context of national cultures that were characterized as being most orientated towards the future, for example, scientists had especially long-term Time Perspectives.

To map ambiguity in Time Perspectives on a global scale the data from the empirical study was extrapolated and reinterpreted using correlations with national scale cultural dimensions for 22 countries. This theoretical study resulted in an indicator of the potential for intercultural cooperation on wicked problems in the water sector worldwide. Brazil, India, and China, countries whose answers to wicked problems in the Water-Energy-Food nexus will be of global importance, are likely to frame problems quite differently from each other but, more importantly, with a longer term vision than recent world powers such as the United States of America, the United Kingdom, and Germany.

The conclusions are important because the time horizons considered in planning and setting research agendas influence what problems are perceived, what questions are asked, and what solutions are sought. An important recommendation is to invest more resources in the framing of problems, goals, and questions. This is particularly important for participatory planning and transdisciplinary research where the diversity in Time Perspectives is greatest. Insight into Time Perspectives can be used in planning to define milestones with temporal targets that are most likely to motivate the relevant actors. It is also useful for directing strategic horizon-scanning activities. For setting research agendas it is important to match the Time Perspectives of those who prioritize the questions with the purpose of the research.

Chapter 1

Introduction

Andrew Segrave

This introduction consists of three parts: (1) a background section; (2) a problem statement and research questions, and (3) an outline of the structure of this book. The background section invokes the recognizable and appropriate phrase of a traditional story: "Once upon a time, in a distant land ...". As later described in Chapter 2, the research was conducted from the outset assuming ontological realism as well as epistemological relativism. This means it is important to reveal some history about relevant personal experiences and thoughts. The background section begins in Australia during the late 1980's and early 1990's where a clash of cultures exposed fascinating diversity in how time can be experienced, conceived, and dealt with. As is often the case with memories, this story makes associative leaps from one recollection to the next.

Ideas concerning the collapse of formerly magnificent societies are explored before zooming in to the contemporary topic of Climate Change. Both of these subjects, which have been recurring themes in popular media since the turn of the millennium, can be seen as signs of how the future is envisaged at present. They are also both relevant to the activities that this research project aimed to shed light on: planning in the water sector and the setting of water research agendas. Water security is generally considered to be one of the most vital ingredients for the wellbeing of a society and Climate Change is seen as a major threat for future water security. This threat includes disasters related to flooding and droughts as well as jeopardizing access to safe drinking water and water for agriculture. Water planning and research can be seen as investments in the future with a purpose of safeguarding water security against such threats. These activities are introduced to conclude the background section. It is from this net of nebulous notions that a problem statement and research questions emerged. These ideas are firstly formulated as a problem and then as an hypothesis before being distilled into six specific questions.

1.1 BACKGROUND

The ideas and experiences that gradually came together to inspire and direct this research project are diverse and may, at first glance, appear to be unrelated. The common thread in these ideas is curiosity and confusion about 'time' and 'the future' as concepts. The conjecture about the issues and questions outlined in this Background section might best be described as the everyday thoughts of a wandering mind. In retrospect, however, these thoughts gradually paved the way influencing the focus of the study and the line of inquiry. Since these ideas were not derived using scientific methods, by any stretch of the imagination, this background section is meant to show from where the hypotheses originated.

1.1.1 Living in different timescapes

Growing up in Australia there were some conspicuous and curious contrarieties between the timescales used in the stories told by authorities about different facets of life. There was, of course, daily life in which some people prided themselves on being punctual, and worried about being late, whereas others didn't seem to give clock time a second thought. But the differences alluded to here go beyond individual idiosyncrasies and involve cultural customs. In stories told via mass media (e.g. the daily News) about changes in the social, political and economic arena, time was measured in 'days' and exceedingly slow processes were seen to take months. Stories emanating from scientists and more significantly, those of the Aboriginal Australians, reflected an entirely different relationship with time. School education during early childhood taught that, due to the lack of tectonic activity, Australia was the oldest continent on Earth having more than half its rock surface dated older than 600 million years. Furthermore, Aboriginal Australians were reported to have one of the oldest continuous cultures on the planet having left Africa up to 75,000 years ago. In contrast with this emphasis, during the bicentennial celebration of the 'first fleet' arrival from Great Britain, Australian children were given memorial coins and released hundreds of balloons to mark a mere 200 years. They heard stories and witnessed re-enactments seeming to imply that little was worth mentioning prior to 1788. While these inconsistencies were confusing to a school child they ignited wonderful curiosity in the author about cultures and time. This curiosity flourished following the bicentenary celebrations as public discussions about indigenous issues became increasingly prevalent and passionate, culminating in the Native Title Act (1993) which recognized the land rights of some indigenous people.

Listening to contemporary Australian stories alongside the re-told Aboriginal Australian stories left the impression they may have been conceived in quite different 'worlds of time'. For example, the scale of time in which the Aboriginal Australian stories unfold seemed fantastically vast. School lessons in social studies would mostly recount events like the Eureka Rebellion of 1854 during the gold rush of the 1850's and 1860's, while school camps commonly involved bushwalks to visit Aboriginal Heritage sites, like Bunjil's shelter. Bunjil is the creator and spiritual leader of the people from what is now called Central Victoria. After visiting his shelter an Elder told the story of how the people had made the sea angry, by catching fish during the spawning season, causing it to rise up covering the plains between the mainland and what is now called Tasmania. The people asked Bunjil to stop the sea from rising so that it wouldn't engulf the entire land and he did so under the condition they promise to respect his laws. From the perspective of western science this Bass Strait was formed at the end of the Pleistocene, about 13,000 years ago. The Elder had told a story about a plain, which was allegedly good kangaroo hunting ground, that hadn't existed for more than ten thousand years. This single anecdote epitomizes further remarkable stories. Hopefully it sufficiently illustrates the long reach back in time of Aboriginal Australian perspectives.

Contemplation of the cross-cultural differences in time perception naturally led to a search for possible signs of these differences in the artefacts of present and past cultures. Impressive buildings such as the Roman Pantheon, built in 126 AD, and the Maison Carrée, from 16 BC built in Nîmes, were among the first things that came to mind. But Australian Aboriginal people were nomadic and did not build cities or use what westerners recognise as advanced technologies - they were socially and spiritually sophisticated. Historical buildings, although enchantingly exotic for someone who grew up in a city where the oldest standing building dates back to 1842 AD, were tangible and more easily recognizable artefacts of cultures than stories, rituals and languages. In hindsight the latter may provide much better grounds for investigating the effects of time perception, but the architectural artefacts of old civilizations, with the Egyptian pyramids as prime example, were and remain a constant source of wonder. Visiting Egypt during this research project and gazing at the deep blue vault of a tomb with the sky goddess Nut arched out over the earth, her body dotted with golden stars and the pathways of heavenly bodies, the deeper significance of time perception became clear. Nut protected people from chaos, safeguarded the dead as they entered the afterlife, and was commonly accompanied by the precious agricultural calendar. The taming of time was certainly essential to Egyptian civilization.

Egyptian fears about death and hopes about an afterlife motivated them to build structures that are still standing after more than 4500 years. On the other hand it is conceivable that the nomadic lifestyle of Aboriginal Australians, with so much energy invested in spiritual and socio-ecological relationships, actually represents a time perspective period longer than that of the Egyptians. Anecdotally, the practices of the former seem to involve waiting, contemplating, and working with time, while the latter involves rushing, structuring, and managing. Part of the inspiration for this book lies in questions about this diversity in cultures, practices, and investments and the idea that different beliefs about time may be at the source of some of this diversity. In any case, both the Aboriginal Australian and ancient Egyptian extremes are quite unlike modern societies and they seem to reflect distinctive relationships with time. A related phenomenon observed while travelling in Egypt is physical proof of societal regression: The earliest pyramids are structurally superior to later ones and some of the most decrepit ruins are of the most recent imitations. Art and medicine also became less refined during certain periods of ancient Egyptian history such as the first Intermediate Period, around 2181-2055 BC, when the society seemed to forget or lose some knowledge and skills. If it happened to them then it could happen to us, so what caused it? One popular physical explanation is the 4.2 kiloyear event, which was an extraordinarily dry period that started around 2200 BC and may have triggered the collapse of both the Akkadian Empire in Mesopotamia and the Old Kingdom in Egypt (Gibbons, 1993).

Societal collapse is explored further in the next section, but let us first posit that climatic events and scarcity of natural resources can cause, or at least catalyse, the downfall of civilizations. Could patterns exist in how cultures dealt with natural resources that stem from their ideas about time and, if so, do some time perceptions coincide with sustainable use of resources whereas other perceptions of time cohabit with self-undermining tendencies? Curiosity about such questions soon tied in with growing fears about the long term future of Australia. In the politics surrounding macro and micro economics the decisions appeared to sacrifice irreplaceable resources for meagre short term gains. Australia is the world's leading coal exporter and also one of the top three suppliers worldwide of zinc, gold, bauxite, nickel, uranium, and iron ore. Most of these raw materials are simply extracted and exported to be refined and processed in other countries, such as Japan and China (Pink, 2012). Rather than investing in more value-adding industries or long-term strategies, the dominance of the short-term seems only to have grown in recent years with hundreds of thousands of hectares of unmined land being sold to foreign companies, like Chinese state-owned Shenhua Watermark Coal. Similarly, large areas of prime agricultural land have also been sold to overseas companies rather than the raw or refined produce, which require a greater (temporal) investment. It is almost surreal to envisage the tens of thousands of FIFOs (Fly-in Fly-out workers in Western Australia) in the most transitory of positions and having such a short-term commitment to a location while leaving such lasting marks on the land (Storey, 2001).

In addition to mining and agriculture, the temporal perspective of Australian political and economic decisions during the 1990's might also have been reflected in forestry practices. While growing up in Southern Victoria the practice of clear-fell logging was devastating rare, cool temperate rainforests. The timber was exported as woodchips for making tissues. This occurred in important water supply catchments for several towns in the region. It is quite difficult to understand the motives for actions such as these when considered from a long-term perspective. The old-growth and complex structure of the forests would be lost and replaced by softwood plantations or new growth that lack biodiversity, water yield and (through a child's eye) beautiful places for bushwalking. Childhood experiences such as this were at the heart of the choice to study subjects related to how people deal with the Earth's systems that support them. Youthful frustration has turned into a deep desire to understand why different groups of people act as they do. Some loggers were genuinely concerned about losing their 'long' family heritage, reflecting on the brevity of their history in comparison with that of the trees they were felling or the land values of the Aboriginal people whose families had enjoyed and managed the trees for many generations. Where did this sincere yet misplaced outlook originate?

A conviction about time perception having far-reaching influences on behaviour also grew through the experience living in two different cultures, Australia and the Netherlands, each for more than a decade. Also working in a Tibetan settlement in India for half a year provided striking insights into some possible consequences of a lack of long-term commitment to a 'location'. It seems natural that people need to have hopes about the future to be motivated in the present. Culturally embedded temporal orientations have been shown to influence the way people deal with natural resources (e.g. Wood, 2008). Other researchers have turned their attention to the accelerating tempo of human activity (e.g. Scheuerman, 2004) and its influence on perceptions of time. In Victoria, Australia, the subversion of Aboriginal ways of conceptualizing and dealing with time has been reported by Nanni (2011) in relation to the manner in which it played a part in colonization. But the early background for this research project clearly was not a review of literature – that came later. The main inspiration came from a profound desire to understand the differences between societies concerning the timescales they perceive and live to. This is an interest obviously shared with Danny Hillis, founder of the Long Now Foundation, who is currently building a magnificent 10,000 year clock to make such time horizons more appreciable. Although the scope and method are obviously different, this research project is based on similar intentions. Here, the aim is also to reflect on and refine the way people think about, and deal with, time and the future by explicating the existing ideas and activities and perhaps even offering alternatives.

1.1.2 Fascination about fears of societal collapse

Around the turn of the millennium there seemed to be increased concern, or at least awareness, about the ephemeral nature of human societies and perhaps even of our species. People appeared to be worried about the future. Barring a handful of eccentrics, like those who support the Voluntary Human Extinction Movement (VHEMT), it is assumed that most people tend to support the continuation of the human race and fear societal collapse. Albeit a debatable proxy, according to Wikipedia proportionally more apocalyptic feature-length films were produced between 2010 and 2013 than in each of the previous decades. Nibiru, a mythical planet, was meant to disrupt Earth's orbit in 2003 but after this botched Armageddon the date was moved forward. Daniel Pinchbeck's (2006) book "2012: The Return of Quetzalcoatl" gave a mixed bag of arguments, including the Mayan calendar, nominating 2012 as the new date and Apocalpyto, a film about Mayan civilization, also hit the screens in 2006. From that point onwards there was increasing reference in popular media to potential cataclysmic or transformative events forecast for 2012 that were linked, much to the dismay of Mesoamerican experts and anthropologists, to the end of the Mayan calendar.

NASA did their best to dispel the myths, but towards the end of 2012 doomsday predictions were popping up everywhere from newspapers to the Discovery Channel and on YouTube, Twitter, and Facebook. Sales of survival kits and even bunker space skyrocketed. Could this widespread, albeit irrational, fear about 'the end of the world as we know it' be given in a less ludicrous explanation than New Age beliefs? Besides rousing a gratifying chuckle, these events again sparked real curiosity about possible causes. Perhaps fear of cataclysmic Climate Change can provide some motives (Bellamy and Hulme, 2011). In 2006, around the time when the 2012 phenomenon emerged, Al Gore's documentary film An Inconvenient Truth became a critical and box-office success, later winning two Academy Awards. This film was meant to be part of an education campaign about global warming. The main arguments it presented were that global warming is real, potentially catastrophic, and human-caused. One essential difference between global warming and Mayan doomsday is that Al Gore's story rests on sound science (IPCC, 2013). Besides this, the time frames for global warming are notoriously nebulous: we foresee catastrophe, but when? If there is one thing people are more afraid of than doomsday, it's an indefinite doomsday (Ellsberg, 1961). Could it have felt comforting to fix a date for our demise? In any case, the time-horizon of 6 years into the future (2006-2012) was conveniently tame: neither awkwardly close nor unfathomably distant – a romantic tragedy.

While the climate science has been rigorously peer reviewed and confirmed (IPCC, 2013), the public fear incited by the risks presented in the Al Gore's film may have become unconsciously coupled with less rational fears. The Global Financial Crisis of 2008 is also unlikely to have improved the outlooks. These are all speculative theories that could be further investigated elsewhere, but the point is that this research project took place against a backdrop of public anxiety about the future and societal collapse, which may have inspired the research questions to some degree. It definitely did inspire a reading of Collapse: How Societies Choose to Fail or Succeed (Diamond, 2005). A professor of geography and physiology, Diamond explores societal collapses in this book from the perspective of environmental causes and warns that current civilizations need to pay more attention to lessons learned from history. Diamond describes the collapse of the Anasazi, for example, a people with a highly developed society between 1050 and 1125 A.D in what is now northern New Mexico. This collapse can be explained through time scales. The Anasazi adopted specialized solutions that were logical and magnificently successful in the short-term but they led to disastrous problems over the longer term. Some anthropologists and historians warn against adopting such simple explanations for collapse and emphasize the need to consider complexity and the reinforcing effects of feedback loops (e.g. Tainter,1990), but overspecialization and a lack of adaptive capacity are generally not considered to be ingredients for future success. Nor is overexploitation of resources.

Although societal collapse most likely results from a complex mix of interconnected internal and external factors, the Earth's systems upon which societies depend, clearly have a maximum carrying capacity or buffer of resilience within which they can remain in dynamic equilibrium. Crossing these Planetary Limits (Rockström et al., 2009), or Tipping Points (Scheffer, 2009), is generally accepted to be unwise if social stability and environmental sustainability are the objectives. The Club of Rome's Limits to Growth is perhaps the most widely influential example of this message from science to society (Meadows et al., 1972). This idea has also inspired scientists and policy makers to define ambitions, such as sustainable resource management, and consequences that are to be avoided, like irreversible environmental damage. An activity is generally considered sustainable if it can be continued at a certain level into the foreseeable future: indefinitely. But what span of the future can one claim to foresee and how might the limits of the foreseeable be defined? How do ideas about what is foreseeable differ across cultures? How is the perception of time, and specifically the future, related to the most widely held solutions for preventing societal collapse, such as adaptive capacity, resilience, and robustness? Are some societies more vulnerable to collapse because of how they perceive and deal with the future? Some of the most successful civilizations, such as the Mayans, Hindus, and Egyptians, worked with calendars that tracked thousands and even trillions of years (Raju, 2003). Debates about this subject are seemingly unlimited broaching what it means to be human. It is almost a moral duty resting on current industrialized nations to investigate the possible loss of some ancient wisdom preoccupation with the 'short-term' and rendering them concerning a macro-history, a loss reducing them to a more vulnerable to societal collapse (Masini, 2001).

1.1.3 Confusion about the temporal grain of Climate Change

The science of 'deep space-time' is a recent and challenging concept. The Galilean revolution, that is, the realization that Earth is not the centre of the universe, along with the Darwinian revolution, comprehending that the human race is not biologically superior to all other species, preceded one of the greatest shifts in western science: recognizing the vastness of Earth's history within which the period of human habitation is basically negligible. Universal processes unfolding in Deep Time do not account for the limitations of human scale perspectives and 'belief systems'. Johnson (1927) put it succinctly: "Geological time is not money." Dinosaur skeletons are major attractions at museums; not only because of the colossal size and monstrous reputation of some species but also because, after being dominant species for millions of years, they became extinct. This is really mind-boggling. To imagine the Earth during the Mesozoic Era involves some fantasy and, as with societal collapse, the downfall of an entire group of animals is awe-inspiring. What's more, this Mesozoic Era is relatively recent in terms of the Earth's history.

Besides being wonderful, and perhaps humbling to some, the extent of Earth's history is so far beyond the experience of daily life that it is generally only understood in terms of metaphors (Gould, 1987). This can lead to errors in thinking, such as the temporal fallacy caused by assuming that patterns discovered through analysis at a given temporal scale will be the same as those resulting from analysis over a shorter or longer period of time (Wood, 2008). Confusion about the relationships between coarse and fine temporal grains has been particularly obvious in discussions surrounding one of the most prominent environmental issues of the last decade: Climate Change. 'Climate' and 'weather' must be among the most commonly confused concepts in present day media. Surveys of educated laypeople also expose this confusion (Read et al., 1994). One particularly memorable example of confusion about the timescales associated with Climate Change is the advice given by a sceptic who suggested that the general public form their own opinions; advocating that people buy thermometers and measure it for themselves. Unless the public were to cooperate in keeping records for decades such data would provide no evidence to support or debunk Climate Change theories.

A more science-based example of temporal fallacy is the attempt to play down the importance of human-caused global warming over the last century. This was done by highlighting the fact that "study of the geological record of climate reveals many instances of natural changes of a speed and magnitude that would be hazardous to human life and economic wellbeing should they be revisited upon our planet today." (Carter et al., 2007). There is no debate about whether or not inhospitable climatic periods occurred millions of years before humans inhabited the Earth. The fact that the Sun will die in several billion years is no less trivial, but it is also not a particularly urgent threat. There may be some retro crocodiles out there interested in knowing what happened to their long-lost cousins, but for us it is more relevant to reflect on how the undisputed rapid rise in greenhouse gas concentrations over the past century are likely to influence the climate system over the next few centuries: consider Climate Change at a scale that is significant to the human species. This is the scale generally adopted by scientific projects directed at informing policy, such as the United Nations Intergovernmental Panel on Climate Change (IPCC), but even the relatively short temporal scale of centuries poses real problems for translating the outcomes into something that is useful for managers making decisions in terms of the next few years (Dessai et al., 2005).

Since the explosion of public interest in Climate Change around 2006, scientists have been struggling to respond to questions from society by downscaling scenarios, for instance those presented in the IPCC's Fifth Assessment Report (AR5), to provide 'policy relevant' information. In The Netherlands, for example, national research programmes, such as the Climate Changes Spatial Planning Programme and the Knowledge for Climate Research Programme, have been directed at 'climate proofing' the country with adaptation strategies that are grounded in science. These programmes also emphasize a demand-driven approach where government and businesses are involved in defining the problems and designing the solutions. But scaling the information from the IPPC scenarios down to a spatial and temporal resolution that is meaningful for these stakeholders is no mean feat and the results are highly dependent on the interpretive framework used (Buizer et al., 2011). The phrase "think globally, act locally" is thus more easily said than done. Besides fascination and awe regarding Deep Time and geological timescales, one aspect of the background for this book was the possible potential to improve the methods and reasons currently used for selecting time horizons and downscaling to plan for Climate Change. This idea arose while planning for projects in the water sector on both (sub)national and international scales.

1.1.4 Hopes set on adaptive planning

At the 2011 World Economic Forum international leaders recognized that "water security is the gossamer that links together the web of food, energy, climate, economic growth, and human security challenges that the world faces over the next two decades" (Waughray, 2011). This quote highlights the importance of water planning and how fundamental water is to supporting the wellbeing of a society. It is beyond the scope of this book to provide an overview of water planning including its history and the current state of affairs. The background for this book is, however, strongly tied into practical experiences over the past 8 years doing futures research and strategic planning for organizations operating in the water sector, designing projects and methods for this purpose, and advising water supply companies and water boards on related questions. These organizations continue to struggle with a necessary transition that emerged over the past few decades. A 'command and control' approach to planning, directed at robust solutions mostly achieved by engineering redundancies and safety margins into the physical infrastructure, is replaced with a more adaptive approach to achieving sustainability (Van der Brugge et al., 2005). Scientific studies concerning this transition are numerous and diverse in their perspectives and theories (e.g. Pahl-Wostl et al., 2007b).

In one particularly interesting meta-level analysis of the transitions taking place in the water sector, Francois Molle (2008) introduced the idea of 'Nirvana concepts' that represent characteristics of an ideal world. A simple yet astute and illuminating observation made by Molle in his analysis is that these concepts usually take the form of a 'photo-negative' of the real world. Using this idea as an analytical tool revealed the characteristics of the problems with which water planners are currently dealing. According to the theoretical ideals dominating contemporary water management literature, water management and planning should be 'integrated', 'adaptive', and 'participatory' (e.g. Pahl-Wostl et al., 2007a). Table 1.1 shows that the sources of the problems perceived in the water sector can be deduced by defining opposites of these Nirvana concepts.

Table 1.1 Nirvana concepts as photo-negatives of problems perceived.

Nirvana Concept	Sources of Problems Perceived in the Water Sector
Adaptive	Inflexibility in a (rapidly) changing context with an (increasingly) uncertain future.
Integrated	Fragmented, simplistic approaches applied to interrelated, complex systems.
Participatory	Ambiguity (in problem perception) and diverse interests (of stakeholders) are neglected using centralized decision making

The purpose of analysing the planning problems in this way was to develop a better understanding of them and become more capable of advising water organizations on how to deal with them. Taking a wider view to learn from other knowledge domains revealed that the type of problems summarized in Table 1.1 can be more generally classified as Wicked Problems (Rittel and Webber, 1973; Lach et al., 2005). Rittel and Webber originally described ten characteristics of Wicked Problems, as summarized in Table 1.2.

These characteristics of Wicked Problems can also be categorized according to their source: (1) the nature of time and specifically the future; (2) the nature of Earth's physical systems, or (3) the nature of human social systems. The Nirvana concepts, or ideal type responses, that are used in the water sector also seem to apply to the ten characteristics of Wicked Problems when they are grouped in this way (Table 1.2).

It seems quite logical that applying 'systems thinking' to water planning problems on the 'roughest' level leads to definition of a boundary between 'the human' and 'the other' in space with interactions occurring over time. In any case, the three Nirvana concepts that the water sector is striving for do appear to be quite generic and abstract. Global frameworks for action, such as the Millennium Development Goals (including the post-2015 Development Agenda), the World Water Assessment Programme (WWAP) and commitments made at the World Summit on Sustainable Development (WSSD), rely heavily on these Nirvana concepts. One ongoing challenge for these frameworks concerns how to refine the ideal type responses to attend to the intricacies, such as the local sociocultural specifics (e.g. Hiwasaki, 2012), of the particular case at hand. Part of the motivation for this book was the idea that a deeper study of the fundamental sources and categories of problems might yield insights for planners and decision makers that are more useful in practice than Nirvana concepts.

Table 1.2 Sources and ideal type responses for characteristics of Wicked Problems.

Characteristics of Wicked Problems (Rittel and Webber, 1973)	Source	Nirvana Concept
1. There is no definitive formulation of a wicked problem. 2. Wicked problems have no stopping rule. 3. There is no immediate and no ultimate test of a solution to a wicked problem. 4. Every solution to a wicked problem is a "one-shot operation"; because there is no opportunity to learn by trial and error, every attempt counts significantly.	The nature of time and specifically the future	Adaptive
5. Wicked problems do not have an enumerable (or an exhaustively describable) set of potential solutions, nor is there a well-described set of permissible operations that may be incorporated into the plan. 6. Every wicked problem can be considered to be a symptom of another problem.	The nature of Earth's physical systems	Integrated
7. Every wicked problem is essentially unique. 8. The existence of a discrepancy representing a wicked problem can be explained in numerous ways. The choice of explanation determines the nature of the problem's resolution. 9. The planner has no right to be wrong 10. Solutions to wicked problems are not true-or-false, but good or bad.	The nature of human's social systems	Participatory

The struggle that water organizations are currently going through in the transition to a more adaptive approach to planning is happening at the most practical level. Looking at the characteristics of the problems for which an adaptive approach is considered an ideal response (Table 1.2), it can be argued that the source lies in the nature of time and specifically the future. Since time itself cannot be directly perceived, people are only aware of it via perception of motion or the change or constancy of a certain perceivable object relative to others (Benjamin, 1981). Time is thus closely tied up with the notion of change in the most fundamental of ways. Change also introduces a second concept that is strongly related to time; that is, human agency and the indeterminacy of the future. The future is influenced by human actions to some extent but it is also uncertain and cannot be known (Adam and Groves, 2007). One of the everyday challenges faced by planners is to work out ways of dealing with this prospective uncertainty without, for example, (unconsciously) substituting possibilities and statistical probabilities for certainties and assumptions (Van Asselt et al., 2010a and b). Based on years of experience working with water organizations on practical planning projects it can be concluded that 'certainty', 'time', and 'the future' are concepts that planners and managers generally do not discuss or define explicitly. An ethnographic study with planners in the Netherlands supports this claim (Van 't Klooster, 2007). Could existing planning strategies be better understood and improved by gaining insight into how water planners conceptualize and deal with time? The idea that this may be the case was one of the main motives behind the research questions addressed in this book.

1.1.5 Questions about the importance of investing in knowledge

Various reasons can be given for investing in research and innovation. In the official documents communicating the objectives of Horizon 2020, the new EU Framework Programme for Research and Innovation which proposes an €80 billion investment from 2014 to 2020, the main motivations offered are the promise of economic growth and jobs. These documents also reason that investment in research will generate the breakthroughs needed to tackle urgent challenges for society, such as Climate Change, and help deliver or secure quality of life. More philosophical and personal motivations obviously exist for striving to acquire new knowledge and tools. How might these motivations differ across cultures? The background for this book includes specific interest in how research is and could be used to deal with one basic fact of life that is seen to make adaptive planning necessary: the fact that future cannot be known.

Meijer (2007) identified four options for dealing with prospective uncertainty: (1) investing in time; (2) investing in flexibility; (3) investing in robustness, and (4) investing in knowledge. An adaptive approach to planning recognizes the fact that (rapid) change and uncertainty can undermine any claims of future robustness. Aesop's Fable about The Oak and the Reed is perhaps the most well-known case against investing in robustness alone: only the reed survives the storm as evidence of the moral that it is better to bend than to break. The Dutch 'Room for the River Programme', which involves providing floodplains and removing obstacles for the river Rhine, rather than just strengthening the dykes, is a recent example of how this principle has been applied to water planning and management. But even though robustness alone is considered insufficient, or at least inefficient, for dealing with prospective uncertainty it is not always possible to emulate a flexible reed.

An adaptive approach differs from a completely reactive one in that some elements of the system being managed are accepted as being relatively inflexible, such as large waterworks infrastructure, which makes continuous adjustment impractical. So an adaptive system is not founded on outright flexibility or robustness alone but on a (pragmatic) balance between these extremes. Taking this into consideration, a water planner is left with the options of investing in time and investing in knowledge. Time tends to be seen as a constraint that needs to be considered carefully in adaptive planning (e.g. Eberhard et al., 2009), whereas investing in research, innovation, and learning to create new tools and knowledge is seen as part of the approach itself. There is plenty of literature supporting the idea that adaptive capacity is directly increased by investing in knowledge. Theorists argue that learning cycles, with recurrent evaluation of the situation and knowledge generation, are necessary to harmonize social and ecological systems and make them more adaptive and thus sustainable (e.g. Pahl-Wostl et al., 2007b). In addition, knowledge can play an indirect but important role for investments in time.

An investment in time may consist of: (1) stalling by delaying a decision or commitment; (2) being proactive to maximise the period available for intervening and/or minimize the time needed to adapt; (3) designing 'exit routes' or reversibility into an intervention, which can be likened to preparing to 'turn back time' if deemed necessary (Meijer, 2007). The idea of reversibility, or delaying commitment to irreversible interventions until the potential consequences are better understood, is also central to sustainability theories – especially those that focus on the quantification of risks (Nachtnebel, 2002). Research and innovation may be directed at generating insights and tools for maximizing the period available for intervening, minimizing the time needed to adapt, and assessing the potential consequences and reversibility of an intervention. These are three ways in which an investment in knowledge can tie in with an investment in time. So apart from directly supporting an adaptive approach to planning, by providing new insights and tools for making informed decisions, an investment in knowledge can indirectly increase adaptivity via time related outcomes. This idea is closely related to other work the author is engaged in, which provides further background to this book.

As a scientist working on water related questions, the author creates, manages, and works on research projects that are financed and organised in various forms, including the collective research programme of the Dutch water supply companies. This research programme consists of: (1) joint research divided into 10 thematic areas; (2) spearhead research directed at advancing knowledge on a specific topic of interest to one or more water companies, and (3) fundamental, exploratory research. The thematic programmes of the International Water Association (IWA) similarly include both the development of pioneering ideas and more applied research that involves practical demonstration of concepts and technologies in case studies. The (water) science system is heterogeneous and it is changing in some quite fundamental ways (Hessels and Van Lente, 2008). To inform and direct the agenda-setting, management, and coordination of these different types of research, the author studies the workings of knowledge networks and conducts futures research.

The Collective Research Programme of the Dutch water supply companies, besides being most familiar to the author, is an interesting case because it has been designed around participatory, integrated, and adaptive ideals. Although the programme managers are currently striving for a greater level of knowledge co-production, employees from the water supply companies are already regularly involved in designing and evaluating the research projects. This is part of a greater trend towards action research that includes practical, together with more conceptual, abstract types of knowledge (Gibbons et al., 1994). In addition to these participatory aspects, the programme was restructured at the beginning of 2013 to be organised around societal themes instead of scientific disciplines; a more integrated approach to knowledge creation. The science system is also changing in reaction to societal pressures for more practical innovations and decision support information (Rip, 2002). One part of decision support involves the identification of prospective threats and opportunities. These insights are useful for planning timely interventions. In this way the research also makes the sector more adaptive.

There is evidently demand for scientists who can work as the 'eyes and ears' of the water sector by signalling problems before they eventuate and developing potential solutions in advance (Sutherland and Woodroof, 2009). In reference to this ambition, documents describing the purpose of the Collective Research Programme of the Dutch water supply companies generally refer to the 'medium to long-term', although the actual length of this period has not been defined explicitly. One of its 10 thematic areas is also specifically dedicated to futures research; analysing current trends and ideas about the future and signalling potential opportunities, threats, and knowledge needs. This research can be seen as an investment in time; that is, maximizing the period available for interventions by systematically scanning the horizon of what might be considered foreseeable. This book aims to address a question related to that process: what temporal horizon is suitable for generating relevant research? Fantasizing about how the world will be thousands of years from now might be considered inconsequential and unscientific, but what temporal horizon is most fitting for futures research that is meant to help in setting research agendas? And what level of certainty do people associate with this foreseeable future?

Besides increasing the time available for interventions, futures research can also represent an indirect investment in other components of an adaptive approach. By raising prospective issues to be considered in setting research agendas, knowledge and innovations can be generated for increasing flexibility and robustness, thus minimizing the time needed to adapt and creating 'exit routes' or reversibility. This does not change the fact that the future cannot be known, however, and herein lies the crux of a second problem that is part of the background to this book. Since the ideas used for setting research agendas cannot be defined as 'knowledge', the agenda-setting process involves negotiating different values and interests. It would be fascinating to gain insight into how agenda setters conceptualize and deal with the future as a way of better understanding the deeper motives behind the criteria used for prioritizing some research questions over others. This is particularly interesting in the light of recent developments that aim to make water research more 'demand-driven' by increasing the influence that actors from the sector have in setting the agenda.

There seems to be a broader trend, evident on the European scale, where piloting and demonstration components and multi-stakeholder 'action groups' are increasingly included as a prerequisite for acquiring research and innovation funding. This trend has also been noticed in, for example, Brazil (Etzkowitz et al., 2004). Do these actors conceptualize and deal with time differently to conventional scientists and if so how might this influence the nature of the research programmes? A research agenda may be collaboratively formulated by managers, practical workers, and scientists, but do these people have a shared problem perception? Considering the amount of time, money, and thought invested in looking for answers to research questions there generally seems to be disproportionately little invested in selecting which questions deserve attention. Perhaps insight into how different types of people conceptualize time and deal with prospective uncertainty could help to improve the existing systems for setting research agendas. The bold hope behind this statement is that this research project may lead to better ways of generating knowledge to make water systems more sustainable.

1.1.6 The common thread: Time as a topic

Time is one main motief in the snapshots of experiences and ideas that came together to generate the curiosity about how people conceptualize and deal with time that inspired this research project. Benjamin (1981) provided a daunting overview of how time has perplexed and frustrated great thinkers throughout the history of philosophy. Gell (1996) also painted a picture of his peers being 'puzzled' and 'mystified' by the subject. It does take some amount of naivety and/or nerve to tackle time as a topic. Through the experience of study the proportions of the latter required to complete this final book grew astronomically. In describing the background to this research project two activities emerged as being especially relevant to the water sector at present and closely related to time perception. Firstly, to deal with wicked problems, such as Climate Change (Hulme, 2009), water professionals seem to have their hopes set on adaptive, integrated, participatory water management. A closer look at the properties of the problems for which adaptive planning is deemed necessary shows that they derive from the nature of time, and specifically the future. An analysis of the strategies that are advocated for adaptive planning also reveals that an essential part of this approach involves investing in knowledge. So while the main topic of this research project is time perception, the implications and applications of the findings will be specifically discussed for (1) adaptive planning in the water sector, and (2) setting water research agendas.

Models and theories about the conceptualization of time are developed further in the Interpretive Framework, firstly from an ontological position and then in a conceptual framework. In concluding this Background section it is important to highlight that a preliminary literature review supported the central assumptions underpinning the research idea; that is, that people vary both in their concepts of time and how they deal with it. Furthermore, that behaviour is influenced by exactly how time is perceived. Each individual has a Time Perspective consisting of multiple dimensions that are formed by various factors (Jones, 1994). Psychological research has shown that people conceptualize and experience time in fundamentally different ways (Lessing, 1968; Zimbardo and Boyd, 2008), and remarkable differences have also been discovered in cross-cultural studies (Noyes, 1980; Boniecki, 1980; Gell, 1996; Levine, 1997). The way people perceive problems and conceive solutions is clearly influenced by this mental construct. As the psychologists Nuttin and Lens (1985) succinctly stated: "not only the present state of affairs, but also future and past events in the frame of a subject's Time Perspective co-determine behavior." Since people performing various roles in different countries need to cooperate to manage water resources sustainably, insight into the dominant Time Perspectives of different groups is likely to be useful for devising transdisciplinary approaches for coproducing knowledge and designing concerted strategies (Thompson-Klein et al., 2004). These conclusions supported advancement to the next step of defining the problem statement and research questions more specifically.

1.2 PROBLEM STATEMENT AND RESEARCH QUESTIONS

The research problem was formulated as follows:

> *The decisions people make, and the actions they take, depend on how they conceptualize and experience time but this fundamental and influential factor is seldom acknowledged, little understood, and rarely considered explicitly in planning; be that for the material systems or the knowledge systems in the water sector.*

This problem is a small yet potentially vital part of a greater and infinitely more poignant problem; that is, that millions of people die each year due to a lack of access to clean water, sanitation and water related environmental catastrophes such as floods and droughts. The basic assumption behind this research project was that planning, in the broadest sense of the word, needs to be effective to change this situation and that to be effective, planning must take the time perspectives of different groups of people into account. The preliminary problem statement can also be framed in terms of a hypothesis:

> *There is likely to be significant diversity in how people working in the water sector worldwide conceptualize and deal with time, which is an important source of ambiguity in how problems are framed and a fundamental reason for the existing shortcomings in how solutions are sought.*

The aim of the research is thus to describe and typify how people working in the water sector worldwide conceptualize and deal with time and to map the ambiguity in time perspectives by characterizing the similarities and differences across groups. A set of six sequential research questions was defined in support of achieving this goal. These preliminary questions were further refined and specified during the process of answering them, as seen in subsequent chapters of this book:

1. Which aspects of an individual's Time Perspective have influence on the decisions they make and the actions they take?
2. How do the relevant aspects of an individual's Time Perspective relate to each other?
3. How can the relevant aspects of an individual's Time Perspective be characterized or measured in an integrated fashion?
4. To what degree are the Time Perspectives of people working in different professions throughout the water sector worldwide similar and/or different?
5. Which similarities and differences in the Time Perspectives of people working in the water sector worldwide are likely to be particularly important for water planning and setting research agendas?
6. Once similarities and differences in Time Perspectives are acknowledged and understood, how can this understanding be used to improve water planning and the setting of research agendas at different scales?

1.3 STRUCTURE OF THE BOOK

Six main chapters comprise this book and are sequenced according to the research questions defined in the previous section. In Chapter 1 the introduction provides a background section painting a picture of the circumstances within which the research questions arose. This assemblage of ideas is then refined in the following section in which the preliminary problem statement and research questions are defined. The second chapter describes the interpretive framework that was developed and used to define, analyse, and interpret the objects of study in a systematic fashion. That chapter includes a new model of time perspective and definitions of the main concepts that were used. In Chapter 3 a new multi-measure method is described, which was designed to characterize the aspects of an individual's time perspective that are most relevant for planning. Details about the assumptions and decisions made in developing the method are included with some empirical results to show the practical consequences together with the selected theoretical approach.

Chapter 4 is a slightly adapted version of a paper that was published in the journal Technological Forecasting and Social Change, which reports on the results of more than 300 one-on-one interviews held with Managers, Practical Workers, and Scientists in the Netherlands, Ghana, Brazil, or Japan. Statistical methods are used, together with anecdotal evidence from the experience of doing the interviews, to identify similarities and differences across groups. That chapter constitutes the empirical basis for the book. In the subsequent Chapter 5 data from the empirical study is extrapolated and reinterpreted to make a map of international ambiguity in Time Perspectives. This chapter is also a slightly adapted version of a paper submitted to a peer reviewed journal. The main result of this theoretical study is an indicator of the potential for intercultural cooperation on wicked problems in water sectors worldwide. In the final chapter the implications and applications of the findings are explored in more detail, including recommendations for water planning and setting research agendas.

Chapter 2

Interpretive framework

Andrew Segrave

The research project documented in this doctoral book embraces polymathy: learning and using knowledge that is not restricted to one field of study. The global science system is under increasing pressure to help solve complex problems for humanity and in response there is a call for research that makes use of all thinkable knowledge without being blinkered by disciplinary boundaries (Van Vierssen, 2012). Starting out as a 'Jack of all trades, master of none', the problem statement and research questions defined in the Introduction (Chapter 1) were initially based on a relatively shallow understanding of the subject matter. The proverbial 'guiding star' was a clear societal problem; namely, the overexploitation of freshwater resources and alarming shortages on a global scale under the current planning, management, and knowledge systems, and an idea about the importance of ambiguity surrounding the perception of time therein. The chosen approach firstly explicated the research questions before identifying and then applying suitable methods from relevant disciplines to generate insights and provide answers from various perspectives.

Desk research was next undertaken with a view to specifying the dependent and independent variables and selecting appropriate existing methods for characterizing them. This literature review and deeper reflection on the task pointed to the seriously difficult challenge of defining variables that depend on concepts such as 'time'. Concepts of time and the language used to define them are dealt with so differently by researchers from different cultures, disciplines and philosophical doctrines that the interpretive framework must firstly be made explicit. This meant taking several steps backward to gain a deeper understanding of the subject matter before again moving forward. The ontological and epistemological stances taken by the author from the outset of the research project strongly influenced the review of existing methods; the decision to develop a new method for collecting empirical evidence; the choices made in designing this method, and the interpretation of the results. Therefore an explanation of these viewpoints is of fundamental practical importance and not just of philosophical interest (Mitroff and Turoff, 1975; Guba and Lincoln, 2005).

Following this line of reasoning, the chapter begins with an outline of the ontological and epistemological grounds that were established to underpin the method. Next the conceptual framework is specified and variables defined which were needed to answer the research questions. In theory, these are the main starting points that inform and influence any process of scientific inquiry (Crotty, 1998). These considerations also resulted in criteria against which existing methods could be tested. The chapter concludes with a summary of variables and criteria for the method. These conclusions were used as guidelines for adapting and integrating suitable elements in existing methods into the new multi-measure method that is described in Chapter 3.

2.1 ONTOLOGICAL POSITION AND MODEL OF TIME PERSPECTIVE

Some explanation of the ontological foundations upon which this research project rests is necessary to understanding the next sections. It is, however, beyond the scope of this project to fully explicate an ontological position. The following synthesis is thus limited to the views that are most relevant to the concept of time. Philosophical positions on the nature of time differ significantly (Adam, 2004). Between the extremes of physicalism (space-like time) and phenomenology (stream-of-experience time) one could map a broad spectrum of stances. The British idealist J.M.E McTaggart identified two dominant and supposedly opposed positions: A-theory and B-theory (Gell, 1996). The position taken in this study includes elements of both theories. Future events are considered to be ontologically different to past events, which is a position attributed to A-theorists. But contrary to the A-theory the present is seen as a subjective state and not a characteristic of real time. Rather than further listing such statements independently, an integrated model was developed using explicit ontological starting points.

More than other philosophical approaches, Critical Realism offers an ontological stance relatively consistent with that taken in this study. The most comprehensive and systematically derived position within this tradition is that of Roy Bhaskar (Carolan, 2005). Bhaskar described reality in strata, beginning at the "real" world of causal powers and deep structures. The next layer is the "actual" world of events that emerge from the deepest level, and the outermost layer is the "empirical" world consisting of that which is observed or experienced (Bhaskar, 1978). Inquiry paradigms that fit this ontological stance accept as true that a mind-independent reality exists (ontological realism), but consider the formative influence of perception and cognition paramount (epistemological relativism). With these ideas in mind various existing models of Time Perspective, such as Monks' (1967) General Model and Katsumata's (2007) Ribbon Model, were compared and analysed.

Husserl (1928) published a model of internal time-consciousness that was adopted and adapted to form the basis for this research project. According to this model, the psychological present is an episode containing various retentions and protentions with temporal duration. Gell (1996) encapsulated this model in what he called its logically prior, but perceptually inaccessible B-series context. Figure 2.1 depicts a modified version of this model, which was adapted to include: (1) Bhaskar's (1978) strata of reality; and, (2) Nuttin's (1964) definition of the future as the time quality of the goal object. These two foundational ideas are explained in more detail in section 2.3: Conceptual Framework. Husserl's model is also broadened to include images of the future and memories of the past, as did Katsumata (2007).

Bhaskar's strata of reality are interpreted as follows: Time Perspective (stratum 3) is a cognitive representation of time based on empirical experience and premises regarding actual time (stratum 2), which emerges from reality (stratum 1). Figure 2.1 only represents the simplest relationships between the essential dimensions. Any of the elements included in stratum 3 may be excluded if the individual perceiver does not consider them to exist. A devoted positivist and believer in B-series time, for example, may only consider the 'V-W-X-Y-Z' series to be real, which would greatly simplify the model. All spatial relationships, which symbolize durations, are also variable. The physical experience of distance has been shown to underlie people's mental representations of duration (Casasanto and Boroditsky, 2008), and various existing methods also employ space as metaphor for time (e.g. Cohen et al., 1954; Cottle and Pleck, 1969; Rappaport et al., 1985). The assumption that distance and duration have analogous mental representations is fundamental to defining temporal extension visually.

Figure 2.1 represents a reference point for the method that is described in the next section by illustrating the relationships between the most relevant dimensions of Time Perspective. The starting point is a given

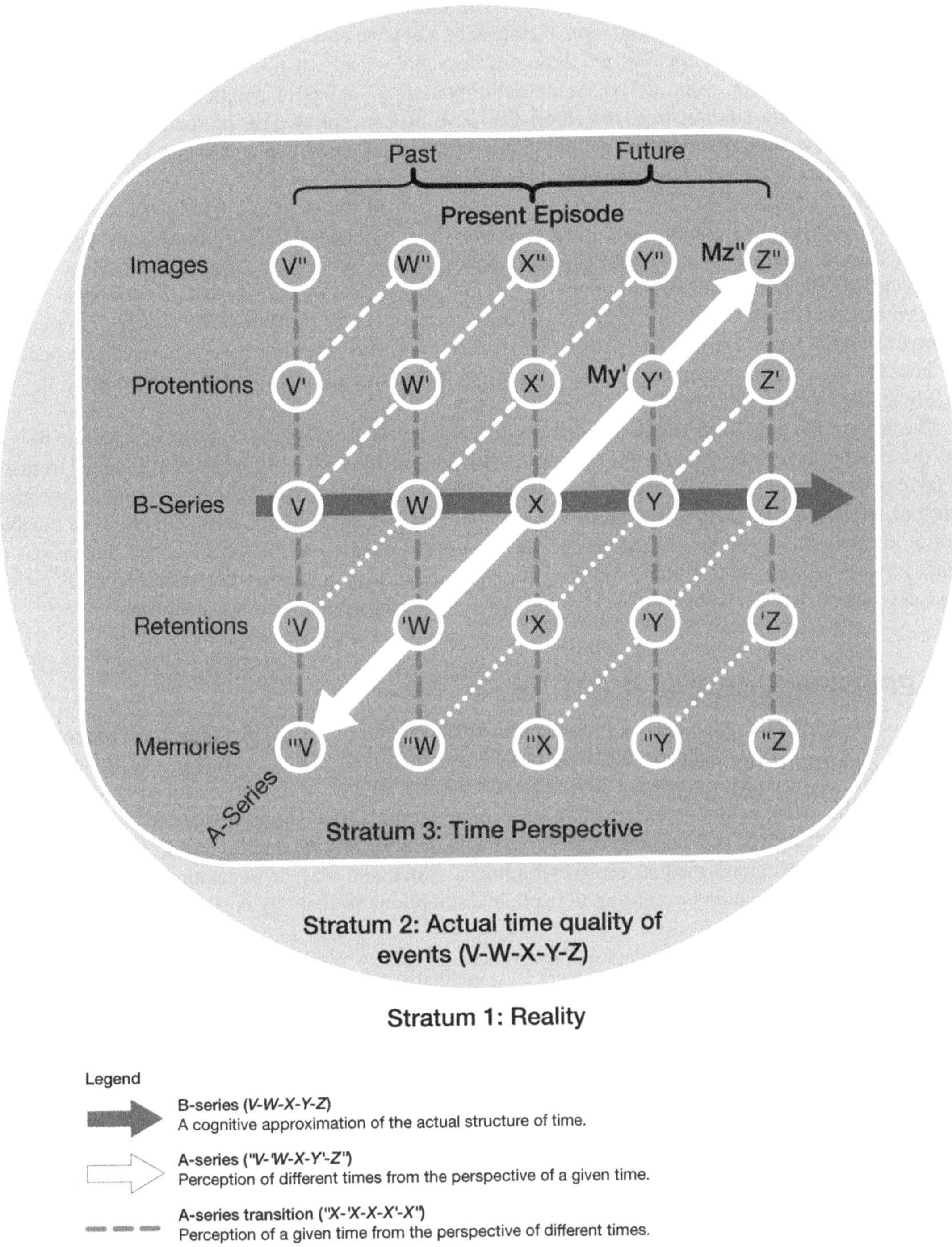

Figure 2.1 Model of Time Perspective (all spatial relationships are variable).

moment 'X' within a subjective episode 'W-Y'. The focus is on the episode because our brains parse experience into episodes that have duration. Retrieval of a bygone episode, conception of a future episode, and response to the present episode, are three neuropsychologically distinct processes (Kinsbourne and Hicks, 1990; Kahneman et al., 2004). Some anthropological studies of people's language and rituals suggest that there are cultures (e.g. the Hopi) that have no concept of past, present, or future (Levine, 1997). But a psychological division of time zones is assumed, even if this division is not universally identical or apparent in every language or culture (Doob, 1971).

The present episode ('W-Y') is subjective. It forms part of the A-series, which includes retentions ('W) and memories ("V) of past times and protentions (Y') and images (Z") of future times. One helpful metaphor for this A-series is a string of pearls: episodes memorized in chronological order (Hancock and Shahnami, 2010). Each pearl is essentially unique and only exists in the mind of an individual. These recalled and envisioned episodes are defined relative to the B-series (V-W-X-Y-Z), which is a cognitive approximation of the actual structure of time. The B-series may be (super)linear or (super)cyclic depending on how the individual conceptualizes the structure of time. Visually, this would mean looping the flat Figure 2.1 into a cylinder or twisting it to make a helix.

The order of the time zones and directionality of the B-series is also variable. Time may be seen to flow in any direction: from left to right, as in most western cultures; from right to left, for speakers of Hebrew; from top to bottom, for Mandarin speakers, and even from east to west, for some wonderfully well oriented Aboriginal Australians (Boroditsky, 2011). The layout of the model depicted in Figure 2.1 must, therefore, be flexible. But to illustrate the assumptions made about (the relationships between) the most relevant dimensions of Time Perspective, and create a framework for structuring the inquiry, a flat, left-to-right version is depicted; since this book is written in English and two-dimensional models are generally less confusing.

2.2 EPISTEMOLOGICAL POSITION

To establish clear starting points for the method, and before defining the foundational concepts more precisely, it is necessary to explain the more basic epistemological premises. This means answering the following questions: What knowledge can be gained about Time Perspectives through scientific research and what is the nature of this knowledge? In keeping with critical realism, it is assumed that ontological questions cannot always be transposed into epistemological terms (Patomäki, 2006). The challenge of establishing an appropriate method involves finding a systematic way of generating new information, knowledge and understanding according to explicit assumptions so that others can comprehend who did what when, how, and why.

The reliability of a method is understood in terms of reflexivity, rather than absolute replicability. This is why the author's assumptions and decisions are described explicitly from the outset (Seale, 1999). Correspondingly, validity is seen as a subjective evaluation of how well the empirical estimates the actual. Judgmental rationality is used in an attempt to transcend the empirical level and to choose between competing explanations of the actual. In other words, different theories about actual things are not just different because of the assumptions they are based on and/or the perspectives of the theorists. There is a theory-independent world that can be explained to a better or worse degree by competing theories. And rational thought can be used to judge which theory is better. In this way critical realism reconciles (1) ontological realism, (2) epistemological relativism and (3) judgmental rationality (Al-Amoudi and Willmott, 2011). Since all knowledge generated by this approach is mediated by historically and culturally partial means of analysis, however, the conclusions are tentative and transient. An important mark of quality in any conclusions is that all transcendent reasoning is grounded in immanent experience (Husserl, 1929).

2.2.1 The example of questionnaires versus interviews

To understand how the epistemological position influenced the research, the following example may be helpful. While designing the method, some scientists questioned the need to interview people personally and claimed that all of the variables of interest might better be measured with good questionnaires - the factorial structure of which can be tested using statistical methods. But such an approach does not fit within the epistemological position because it overemphasizes the significance of quantitative data relative to other empirical input for developing knowledge and understanding, such as listening to informal comments and watching body language. For a similar reason it was considered essential that the researcher be immersed in the different cultures rather than assume that statistical evaluation of construct equivalence would suffice in grounding cross-cultural comparisons. The contextual theory-dependence of knowledge production cannot be eliminated by methodological prowess (Al-Amoudi and Willmott, 2011), and the more types of knowledge incorporated the better.

2.3 CONCEPTUAL FRAMEWORK

Concepts forming the basis for a more specific definition of the variables and research questions need clarification though key concepts identity and perception involve theories that are contested. All of the concepts build on the model of Time Perspective (Figure 2.1) and the ontological and epistemological foundations explained in the previous sections. Assumptions made here result in criteria for the research method. Next, it is necessary to describe the theoretical background of the two main concepts that form a basis for specifying the variables. Motivational Space and Time Perspective. Finally, the concepts used to differentiate between groups and define a sample population are described. The method also had to meet certain criteria to work with these groups. The variables and criteria for the method are summarized at the end of this section.

2.3.1 Identity

While recognizing the differences between individuals in their Time Perspectives, differences may also exist between identities within an individual. Various social psychologists have argued that people assume various identities according to the role they are performing (Goffman, 1959; Schlenker, 1985; Ting-Toomey, 1993). An individual may alternate between being mother, sister, sportsperson, and manager - all in one day. By a process of 'identity negotiation' individuals establish mutual expectations of one another to form groups (Swann, 2009). Such flexibility in identity is also likely to influence one's Time Perspective. The method must therefore limit each individual's identity to their Professional Role (criterion 1).

2.3.2 Perception

A belief is made real through its consequences, but this does not make the object of that belief real. Human perception is fraught with biases (e.g. Azariadis, 1981; Pregernig, 2002; Sanna and Schwarz, 2003; Busby et al., 2009). Decisions are seldom rational and the decision making process is influenced by conditioning, empathy, hyperbolic discounting, emotions, and impulsiveness (Ainslie, 2005). Considering the ontological and epistemological viewpoints described in the previous section, it was decided not to use methods that assess the realism of an individual's views, even though previous researchers have identified realism as one of the dimensions of Time Perspective (e.g. Lewin, 1951). First person reports of reasons for acting were accepted as explanatory (D'Oro, 2013).

A second dimension of perception that was necessary to consider was the ability to intentionally misrepresent one's views. Groups of individuals generally have a value system associated with how they perceive their role, which also applies to Time Perspective. Foresters, for example, are expected to look ahead and plan for long-range goals (Hoogstra, 2008). There are other reasons for people giving socially acceptable answers; for example to avoid appearing unintelligent or irresponsible. Western researchers report difficultly in drawing candid answers from people from eastern cultures (Bian, 1994). People are also influenced by their inferences about the aim of the research (Norenzayan and Schwarz, 1999). The method must thus account for socially acceptable responses (criterion 2). Pre-emptive control using the sequencing of questions and methods triangulation was preferred over post-study statistical corrections based on social desirability scale scores (Ellingson et al., 1999).

2.3.3 Motivational Space

If a concept influences how someone thinks or acts (it is causally efficacious) then it is considered to exist (Walton, 2008; D'Oro, 2013). This is an important starting point for defining Motivational Objects and subsequently Motivational Space: Motivational Objects are the things that motivate someone to make decisions and act (what). Motivational Space is where these objects are located in space and time (where and when). It is assumed that both of these concepts represent actual psychological traits, as have various psychologists (Wallace, 1956; Piaget, 1969; Nuttin, 1984; Block, 1990). In Chapter 4 these concepts are discussed in more detail, specifically in relation to the framing of problems, and developed into a model of Problem Space. Figure 2.1 illustrates how Time Perspective is characterized using the temporal dimension of two Motivational Objects (My' and Mz''). The concept of Motivational Space formed a foundational framework within which the variables were defined, together with the concept of Time Perspective.

2.3.4 Time Perspective

Time Perspective is defined as a context dependent, multidimensional construct that includes all of the different views on time held by an individual. Individuals form groups that have collective views on time, known as Time Frames. The Time Frame of a group can be characterized by the Time Perspectives of the individuals it comprises (Noyes, 1980), but it can also be seen as a cultural characteristic with its own emerging properties (Levine, 1997; Hofstede et al., 2010). The individual was selected as the most clearly autonomous entity for defining the typical Time Perspective of groups, but the concept of 'self' is also recognized as a relational phenomenon (Heine, 2001). Considering the definition of 'identity' provided above, this made it especially important to characterize each individual's Time Perspective while situated in the context of the group in question. But even with a clear focus on an individual in a specific context, Time Perspective remains a complex and elusive variable.

Zimbardo and Boyd (2008) claim that no two individual's attitudes toward time are identical. Previous empirical research also showed that there is limited overlap between the various markers of Time Perspective and that the outcomes tend to be largely a function of the method used (De Volder and Lens, 1982; Gjesme, 1983; Rappaport et al., 1985). Adams and Nettle (2009) found significantly correlated markers of Time Perspective, but the strength of correlations was not strong. One likely reason for the lack of congruity between measures of Time Perspective is that each measure is designed to measure a specific dimension and the different dimensions may vary independently. A person who is biased towards the future, for example, need not have a particularly extensive long-term vision or share with other future-biased individuals a certain concept of how time, including the future, is structured. There is great potential for diversity in Time Perspectives which, considering the aim to uncover patterns on a group level, was a bit disconcerting. This is why it was considered imperative to define the various dimensions relative to each other in a model (Figure 2.1). The ambiguity needed to be restricted, as much as possible, to that across individuals with reference to one specific dimension.

Research was focused on the dimensions of Time Perspective that are likely to have the greatest influence on how people respond to wicked problems, which are assumed to concern forethought and planning (Rittel and Webber, 1973). This assumption is based on previous research which found that motivation for decisions and behaviour is related to the present anticipation of future goals (Polak, 1973; Simons et al., 2004). The 'future' is taken to mean times after a subjective present: that which is yet to be. This comprises both the bodily (from the present episode until the point of imagined death) and the transcendental (from death onwards). More specifically, Nuttin (1964) defined the future as "the time quality of the goal object; the future is our primary motivational space." De Volder and Lens (1982) used this premise to integrate the dimensions of Time Perspective that concern the future into cognitive theories of human motivation. This previous research formed the basis for selecting and defining the dimensions of Time Perspective that would be used to characterize the dependent variable. Lewin (1951) made the distinction between the Range and the Realism of an individual's views on time. Subsequent efforts (e.g. Jones, 1994) identified more dimensions of Time Perspective, including Extension, Density, Valence, Accessibility, Content, and Structural Organisation. Building on the concept of Motivational Space and the model of Time Perspective it was decided to focus on Extension as defined by Motivational Objects:

Motivational Objects: self-defined causally efficacious goals and events (what).
Extension: the span of time into the future that is populated by Motivational Objects (when).

Motivational Objects must influence decisions and actions to be considered real. There has been much criticism of methods that fail to differentiate between cognitive and dynamic aspects; lumping Motivational Objects with ineffectual wishful thinking (Klineberg, 1967; Wohlford and Herrera, 1970; Lessing, 1972). But people have fundamentally different definitions of what is 'foreseeable': to what degree can a belief be justified as true before the object of the belief exists? The term 'foreseeable' implies that the future is essentially unknown and refers to phenomena that can be reasonably anticipated or forecast by applying ordinary reasoning. What is reasonable or ordinary is subjective and part of this research project was directed at investigating determinants for this psychologically and culturally dependent variable. Given the ontological framework and epistemological stance, a third dimension was added to the dependent variable to account for subjective perception of reality:

Certainty: attributed to Motivational Objects by the interviewee (how foreseeable).

Certainty is no measure of realism: If someone believes in fairies and makes decisions and acts based on this belief then the fairies will be used to define that person's Motivational Space. The fact that the object forms the basis for decisions and actions is what was considered important. The dependent variable, Foreseeable Future, is thus represented on Figure 2.1 as the temporal extent (X-Z") of Motivational Objects (Mz") and the Certainty attributed to those objects. The Foreseeable Future is relative to a context that is characterized by other aspects of Time Perspective (Figure 2.1). This context needed to be defined using two dimensions:

Structure: concept of time itself including definition of the time zones.
Orientation: bias towards or against the future relative to the past and the present. In Figure 2.1 orientation is represented by the perceived importance of future (X-Z") relative to present ('W-Y') and past (X-"V).

Other dimensions of Time Perspective, such as Content, Density, etc., were assumed to represent more detailed characteristics of Motivational Objects and were not included as distinct dimensions.

2.3.5 Sample population and groups

Rather than viewing culture as a framework or setting it was defined as a shared quality that is intrinsic to the 'self' (Geertz, 1973; Heine, 2001). Even so, national borders were accepted as rough divisions between cultures without making any assumptions about causality; that is, did cultural differences lead to borders or the other way around? Various studies have compared Time Perspectives between countries (Levine, 1997; Raju, 2003; Hofstede et al., 2010). Based on this earlier research, four countries were proposed for representing the variance across National Cultures: the Netherlands, Ghana, Japan, and Brazil. As regards the level of abstraction of one's Professional Role, from theory to practice, three sample groups were defined:

Scientist: a specialist in one of the physical or natural sciences related to water management.
Manager: someone who controls resources, expenditures, and the direction of (part of) an institution, which in turn directly influences water resources.
Practical worker: someone who physically intervenes in the management of water resources, generating material outcomes with their individual actions.

These roles were chosen for practical reasons; considering the structure of the water sectors in the countries of interest. The option of including additional groups, such as policy makers, was considered but the three groups selected were assumed to represent variation across different Professional Roles sufficiently and organizing interviews with additional groups would be impractical. For developing the method, the primary concern was that it could be used with these groups at least. The research questions defined in Chapter 1 reflect the focus on how similarities and differences in the Time Perspectives of groups influence two activities: (1) water planning, and (2) setting research agendas. The individuals involved in water planning may include Scientists, Managers, Practical Workers, and people fulfilling various other roles from different countries. This is also true for the setting of water research agendas. If an organisation or network of 'water planners' or 'agenda setters' can be specified; that is, different people working on a joint mission, the aim of this research project is to characterize the intra-network diversity in terms of Time Perspective. The hypothesis is that there are differences between individuals in these organisations or functional networks, concerning how they conceptualize and deal with time, which are related to their National Cultures and Professional Roles.

The nature of the sample population and subgroups imposed three main requirements on the method. Firstly, since it was not possible to use pretests and posttests with a single group, the method had to facilitate static group comparison over a short period (criterion 3). National Cultures are considered dynamic, as are the Time Perspectives of individuals. But both these processes are assumed to be relatively slow: Individuals age over years and cultures are unlikely to change overnight (Lessing, 1972; Hancock, 2002; Lachman et al., 2008). Even so, the value of experimental research methodology that involves follow-up research using repeated measures cannot be underestimated. Future studies may build on this research project in such a way. The second criteria related to the sample population results from the fact that the method must be adapted to different languages and cultures to make meaningful cross-cultural comparison possible. It was necessary to work with constructs that are comparable between groups. Preference was thus given to existing methods that have proven useful internationally, particularly in the Netherlands, Ghana, Japan, and Brazil (criterion 4)

2.3.6 Summary of variables

Analysis of the concepts described in the Conceptual Framework resulted in specification of the variables and criteria for the method. The dependent variable, which is called Foreseeable Future, was defined as the temporal Extension of an individual's Motivational Objects and the Certainty attributed to these objects. As described above, the three dimensions that define the dependent variable are a subset of those that constitute a Time Perspective. Foreseeable Future is relative to the Structure and the Orientation of each individual's Time Perspective. The independent variables were defined as National Culture and Professional Role.

2.3.7 Summary of criteria for the method

The Conceptual Framework introduced four criteria for the method:

1. Limit each individual's identity to their Professional Role.
2. Account for untrue socially acceptable responses.
3. Facilitate static group comparison over a short period.
4. Proven useful in different languages and cultures.

Besides these specific criteria, the ontological and epistemological position and the model of Time Perspective set more broad, fundamental requirements for the method. It would need to, for example, produce empirical data while maximizing the formation of other types of knowledge and understanding through experience. On a practical level, to make scientific testing possible, the empirical data obviously needed to characterize the independent and dependent variables as well as a set of standard confounders, including: age, gender, affluence, health, education, religion, some of which have also been found to influence Time Perspective (Levine, 1997; Zimbardo and Boyd, 2008).

Chapter 3

A new method

Andrew Segrave

Various methods exist for characterizing specific aspects of how individuals conceptualize and deal with time. This chapter begins by reflecting on these existing methods but it is purposefully weighted towards explaining a new multi-measure method for characterizing the Foreseeable Future and how this method was developed. The main purpose of this chapter is to explicate the method so that the empirical investigation might be imaginatively replicated by others. It is not, however, considered possible to reproduce exactly any empirical study involving people. As explained in the epistemological position (section 2.2), reliability is understood in terms of reflexivity rather than absolute replicability. Assumptions and decisions made in developing the Foreseeable Future Multi-measure Method are explained separately for each step in the interview schedule together with some empirical results to show the practical consequences together with the selected theoretical approach.

3.1 EXISTING METHODS

As stated in the Conceptual Framework, the individual was selected as the most clearly autonomous entity and window into group culture. The search for suitable existing methods thus centred on the field of psychology, though alternatives like Robert Levine's (1997) pace of life proxies were also considered. More than twenty psychological measures were found that were partly relevant as well as five prior comparisons of methods for measuring different aspects of Time Perspective (Lessing, 1968; De Volder and Lens, 1982; Strathman et al., 1994; Keough et al., 1999; Adams and Nettle, 2009). Adams and Nettle's (2009) study statistically investigated correlations between several different methods, while the other studies involved more qualitative comparisons.

It is beyond the scope of this research project to describe or evaluate in depth the various existing methods. These methods vary with different philosophies and models of time, although the ontological and epistemological grounds were generally unclear or left to unnamed implication. This made it difficult to assess exactly which dimensions of Time Perspective they were designed to characterize. But the primary reason for developing a new method, rather than adopting an existing one, was that no existing method explicitly characterized each of the relevant dimensions of Time Perspective relative to each other based on an explicit model of Time Perspective. Of the existing methods, Zimbardo's Time Perspective Inventory was clearest in this regard but it did not include a measure of Extension, which was central to this research project and the definition of the Foreseeable Future (Chapter 2.3). The pretest analysis also revealed some serious limitations concerning this type of inventory, as described in the section on Step 5 in section 3.2.3 below, which is why this kind of instrument would not be used as a primary measure for any of the dimensions of Time Perspective.

3.2 FORESEEABLE FUTURE MULTI-MEASURE METHOD

In this section descriptions of the theoretical assumptions and decisions that were made in developing the Foreseeable Future Multi-measure Method are, where possible, reported with samples of the empirical data that was generated through applying the method or anecdotes about the experience. This treatment of the theoretical considerations together with the practical consequences is meant to demonstrate the nuances of each step so that they can be understood and imaginatively replicated.

3.2.1 Identify the target sample population

Based on the arguments introduced in section 2.3 it was decided that each individual from the sample population should be interviewed in person and that a single interviewer should conduct all of the interviews together with a local interpreter. The Foreseeable Future Multi-measure Method itself thus facilitates continuous reflection and learning throughout the research project so that any divergences of meaning can be explored and understood. Conducting one-onone interviews would also allow the researcher to sense first-hand whether or not the interviewees understood the abstract subject matter. Alternative methods, like questionnaires, might produce the same theoretically verifiable data but not the same level of understanding in the researcher's mind. After deciding to survey the sample population using one-on-one interviews, a target sample population was defined to get a feeling for the task at hand and to direct the design of the method.

A minimum sample size for avoiding biases and ensuring sufficient statistical power to find significance was determined assuming that the dataset would need to be suitable for multilevel regression analysis. Hox (2010) indicated that more than fifty groups would be needed with at least five individuals per group. Raudenbush and Liu (2000) also found that the number of groups has greater influence on statistical power than the group size once a minimum of five individuals per group has been attained. While understanding the dynamics of the statistical procedures it was decided to be pragmatic and conduct post hoc analysis of significance of the results rather than design the project around these limitations. For various reasons, as explained in the Epistemological Position, more value is attributed to the knowledge gained through experiencing interviews first-hand than to statistical arguments. The practical limitations that come with this approach would make it unfeasible to travel to more than four countries or to interview people from more than three professional roles. The number of groups was thus limited to twelve. Future research should, however, aim to deal with this limitation.

With the intention of producing a representative sample of the variance in Time Perspectives across cultures worldwide, including different roles in the water sector, twelve groups were delineated in the Conceptual Framework. No statistically founded sampling method was used, since the author depended on local contacts to arrange the interviews. Future studies might investigate the possibility of employing more robust sampling techniques but for this research project it was unfeasible. A standard letter of request was written in English, translated, and sent to the main water organisations in each country. This letter requested interviews with practical workers, managers and scientists from the addressed organization and provided definitions of each Professional Role. The research focus was kept secret by using vague sentences such as: "With this research we aim to develop insight into the perspectives of different groups in the water sector." The purpose was to minimize the inferences that interviewees could make about the goals of the project (criterion 2).

A pragmatic approach was also used in determining the number of individuals per group. For financial reasons the author had four weeks in each country to interview as many individuals as possible while maintaining comparable group sizes. Since it was necessary to interview people in the context of their workplace (criterion 1) the weekends could not be used. Taking travelling time into consideration, which would be considerable for a country as large as Brazil, a maximum of four interviews per day and twenty days was assumed. The target sample population was rounded off to three- hundred: twelve groups of twenty-five individuals (Table 3.1). Future studies might investigate the possibility of including more groups. The most efficient way of achieving this is likely to be the addition of extra Professional Roles, such as Policy Makers, but this decision naturally depends on the research questions.

Table 3.1 Target Sample Population

	Practical worker	Manager	Scientist
Netherlands	25	25	25
Brazil	25	25	25
Japan	25	25	25
Ghana	25	25	25

3.2.2 Pre-test considerations

Future studies that use the Foreseeable Future Multi-measure Method should always begin with certain pre-test considerations. Assessment of language translations and concept equivalence, for example, will remain a recurring necessity. Since the method was used for the first time in this research project, however, various extra considerations needed to be taken into account prior to conducting the first interview. In addition to the criteria that stem from the particular demands of the study and the viewpoints of the author, as described in Chapter 2: Interpretive Framework, several checklists and guidelines for proper adaptation of methods for cross-cultural studies were also consulted (e.g. Hambleton et al., 2005). It should be noted that existing measures were adapted in such a way that they cannot be considered equivalent to the originals. The reflexive review of the adaptation and integration of measures in this Chapter is meant to demonstrate the reliability of the new method. Much attention was paid to critical self-awareness throughout the design process. For example, the importance of intersubjective revision of results with members of the sample population during the testing period was emphasized.

The importance of consulting people from within the different cultural groups during the development and translation of the Foreseeable Future Multi-measure Method was emphasized throughout the design process. The author prepared a detailed Interview Schedule that prescribes every step and every cue in the language of the interviewee. The concepts and the language used in this schedule were discussed independently with at least two local translators in each country to identify different interpretations before cross-checking and collectively discussing inconsistencies and ambiguities. Where possible, emigrants living in the Netherlands were also consulted for their first-hand experience with intercultural interpretation. Any future studies that use the Foreseeable Future Multi-measure Method need to include pre-test analysis of these considerations in this way. What follows is reflection on an example of a result from the pre-test analysis conducted for this research project to highlight the type of considerations and to emphasize the importance of this preparatory step.

The example of Japanese motivation

While various translation issues were identified and unravelled for each country, the most significant example involved finding a Japanese equivalent for our definition of 'goal'. The Japanese word 'mokuhyo' refers to various concepts ranging from an 'aim' to what might best be described as a 'guiding principle'. One problem with this concept was that a 'mokuhyo' is often seen as timeless; never to be realized. An alternative translation was found in the word 'kadai', which is generally associated with a certain time frame and translates as 'task' in English. But a 'kadai' is not necessarily related to an individual's view and may be obligatory or assigned by a senior colleague. The link with motivation may thus be questionable. Conversations about these constructs with Japanese interpreters supported previous research showing that the concept of motivation could differ fundamentally between individualistic and collective societies (Hofstede et al., 2010).

People in Ghana, Brazil, and The Netherlands seemed accustomed to having personal goals and priorities whereas Japanese people tended to presume that such matters would be collectively defined by a group. When asked "In your opinion, what are the most important goals that need to be achieved in your field of work?" one Japanese Practical Worker replied "I can't remember all of our goals – can I go fetch our planning document?". To account for these differences between the Japanese and people from the other three countries the focus was set on objects, anticipated by interviewees, which influenced their decisions or actions. This example illustrates the importance of comparing constructs in different cultures. Prior to applying the Foreseeable Future Multi-measure Method for any future research it is necessary to analyse the concepts defined in the Conceptual Framework (section 2.3) and in the interview schedule together with representatives from the populations of interest.

3.2.3 Interview procedure and discussion

The Foreseeable Future Multi-measure Method procedure is prescribed in a detailed Interview Schedule, which is extensive and thus excluded from the main text. Besides dictating instructions in a particular order and manner this schedule is also essential for structuring and controlling language translations. The procedure summarized below was followed meticulously for every interview and each interview was also transcripted using a standard format that may be requested from the author. The questions were printed on separate call-cards, which were first read aloud then handed to the interviewee. To help limit each individual's identity to their Professional Role, the interviews were held at the interviewee's workplace (criterion 1). Every interview began with a check of background information and the gathering of data on standard confounders, such as age, gender, affluence, health, education, and religion. The idea was to take the time to become comfortable with the interviewee by asking some simple questions without mentioning Time Perspective or the specific purpose of the interview (criterion 2).

Ideally, the distributions of confounders are comparable across all groups. But for the empirical test this was not the case. For example, more female scientists were interviewed than female practical workers and more Brazilian females than Ghanaian females. The gender profiles of the sample groups did correspond with the gender profiles of the populations of interest. Even so, the influence of gender obviously needs to be tested using multivariate analysis. This illustrates the importance of collecting data on confounders, including gender, age, affluence, health, education, and religion.

The interview itself was divided into five successive steps that were sequenced to account for untrue socially acceptable responses and to facilitate triangulation (criterion 2). The following description is structured around these five steps. Each step focusses on one specific aspect of Time Perspective, as illustrated in Figure 3.1, which is organized around the Model of Time Perspective depicted in Figure 2.1. Triangulation was made possible by using at least two measures for each of the dimensions of Time Perspective, except for Structure. The Circles Test for Structure (Step 2) itself includes a visual illustration and a textual legend, which is meant to check the meaning.

The five steps of the Foreseeable Future Multi-measure Method generally take 30 to 60 minutes to complete. If an interviewee obviously didn't understand a question then the interviewer explained the meaning and took note of everything that was said. This was also true for any other digressions from the schedule or interesting comments. Each interviewee drew or wrote down their responses to all of the questions and this, together with the interviewer's notes, made up the record of each interview. The interviews were then digitized and the numeric data was tabulated for input into statistical analysis software.

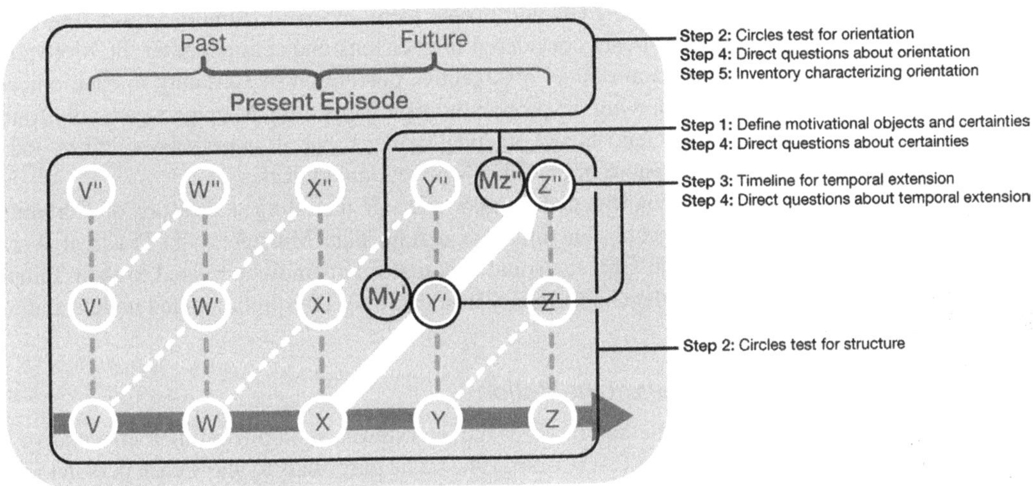

Figure 3.1 Variables and measures associated with the 5 steps of the new method.

Materials

To complete the five steps of the Foreseeable Future Multi-measure Method certain materials are needed. All of these materials were purchased and organized for all four countries before beginning the first interview to make sure that there was no variance in the visual cues given to interviewees from different groups. A ring binder containing one punched plastic pocket per interviewee was prepared for each country. Every plastic pocket held:

- 1 cover page with anonymous interviewee background information
- 1 blank piece of A4 paper
- 1 60 cm long, 5 cm wide, blank strip of adding-machine paper
- 1 printed questionnaire with 5 direct questions and blank spaces
- 1 printed inventory with 30 items and a 5 point Likert scale

The researcher also brought certain standard materials to each interview:

- 1 interview schedule printed in English
- 1 interview schedule printed in the interpreter's language
- Yellow Post-It notepads
- 10 questions, each printed on a separate call-card
- 2 pens

Step 1: Define Motivational Objects and Certainties

The first step involved identifying and recording the interviewee's Motivational Objects, and the Certainty attributed to these objects, before making any reference to time (criterion 2). Techniques that work with personal future events have been widely used in past investigations (e.g. Wallace, 1956; Klineberg, 1967; Lessing, 1972). These existing methods were adapted to satisfy the criteria. Instructions and other cues were minimized and rather than working with a predetermined inventory of Motivational Objects (e.g. De Volder, 1982), interviewees defined their own ones All instructions specifically refer to work related goals and events (criterion 1).

The interviewee wrote each of their goals and events on separate Post-It stickers to allow for independent (re)positioning. All intrinsic and extrinsic goals and events were accepted (Simons et al., 2004) under the condition that they had been consciously considered in decisions and actions. After the Motivational Objects were defined, the interviewees attributed a subjective perception of Certainty to each object by writing a percentage on each Post-It showing the probability of it happening. Managers generally found it easier to write down their goals than scientists and practical workers, but all interviewees understood this step and gave answers. Figure 3.2 is a photo of one such Motivational Object.

The results of the empirical test confirmed the importance of including a measure of Certainty in the method. Scientists, for example, were generally less certain than Managers and Practical Workers about whether or not their Motivational Objects would eventuate. This may be related to their Temporal Extension, for example, which makes data on Certainty invaluable for interpretation and understanding of the results from other measures.

Step 2: Characterize Structure and Orientation

The second step in the multi-measure method involves the interviewee symbolizing their concepts of time using adapted versions of Cottle's Circles Test (Cottle, 1967). The purpose is to allow each person to define a structural context within which the Temporal Extent of their Motivational Objects can be identified; thus

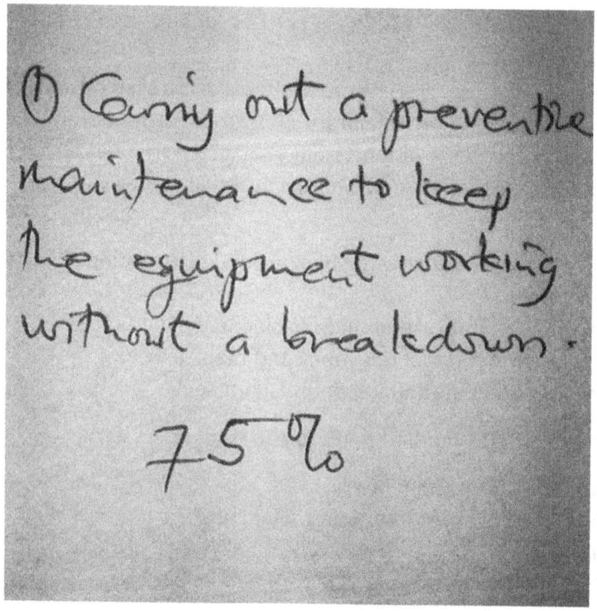

Figure 3.2 A Motivational Object.

allowing for different images of time. The Circles Test has been criticized for its reliability and scoring problems in the past (e.g. D'Alessio et al., 2003), but other studies have used it to generate meaningful results (e.g. Bruno, 1995; Brown and Herring, 1998; Hoogstra, 2008). Besides being simple and quick (criterion 3), the Circles Test minimizes the limitations of textual translations by using visual symbolization (criterion 4). As a visual instrument, the Circles Test was also simple to modify for integration with other measures that use space as a metaphor for time.

Two Circles Tests were used: one represented the interviewee's Orientation to each time zone as regards its subjective importance for their decisions and actions. The second symbolized the Structure of time itself. Separate tests were conducted to avoid confusion between Orientation and Structure in a single test. The empirical results may be seen to support this decision, since the two tests produced different results. On the other hand, interviewees are unlikely to assume that they are being asked the same question twice and would thus be inclined to give different answers.

The first Circles Test involves drawing three circles showing the relative amount of influence that the past, present, and future have on the interviewee's work related decisions and actions: the bigger the circle the more influence. The relative size of the three circles drawn in the second Circles Test represents the amount of time in each time zone: the bigger the circle the longer the duration. Most interviewees understood and completed the first Circles Test with little difficulty, but some took longer to complete the second Circles Test. The explanation that interviewees generally gave for this was that they had never thought about the durations of time zones before. Interviewees were also asked to provide a legend for each Circles Test to verify how they understood the instructions and aid interpretation. The results of the empirical study verified that provision of a legend was indispensable to verifiable interpretation of the drawings and this step also helped the interviewees to clarify their ideas.

The second Circles Test is specifically tailored to represent a structural context within which the temporal extent of the Motivational Objects can be specified in the next step. Interviewees draw the path of time using an arrow between the circle representing the present and that representing the future. This arrow confirms the arrangement of the circles and visually embeds a timeline in the drawing of temporal structure. Figure 3.3 is a photo showing one Ghanaian interviewee's response to the Circles Test for Structure question. This response was the mode for Ghana; 24 of the 79 interviewees pictured time in this way, and also for all interviewees if National Cultures and Professional Roles are lumped.

An analysis of Figure 3.3 reveals that the three time zones are separate (no overlap), that they are arranged from left to right on the page, and that they are sequenced from the past through the present to the future. The size of the circles indicates that the future comprises the most amount of time, followed by the present and then the past. There is also a linear arrow from the present to the future showing the perceived path of time. The significance of these results is discussed in later chapters. In preparation for the analysis, each interviewee's illustration was codified and digitized. For example, the relative size of the time zones was expressed using (in) equality signs and then sorted into ordinal categories:

1. Future absolute largest:
   Pa<Pr<Fu; Pa=Pr<Fu; Pr<Pa<Fu;
2. Future equal largest:
   Pa<Pr=Fu; Pr<Pa=Fu;
3. Other:
   Pa=Pr=Fu; Pr<Fu<Pa; Pr>Fu>Pa
4. Future equal smallest:
   Pa>Pr=Fu; Pr>Pa=Fu;
5. Future absolute smallest:
   Pa>Pr>Fu; Pa=Pr>Fu; Pr>Pa>Fu;

The order of time zones from left to right was also codified and categorized and each illustration was classified using two binary truth conditions: overlapping zones (1; 0) and linear time (1; 0). The Circles Tests were categorized and modes were determined for each group to characterize the context within which Temporal Extension could be defined in terms of Structure and Orientation.

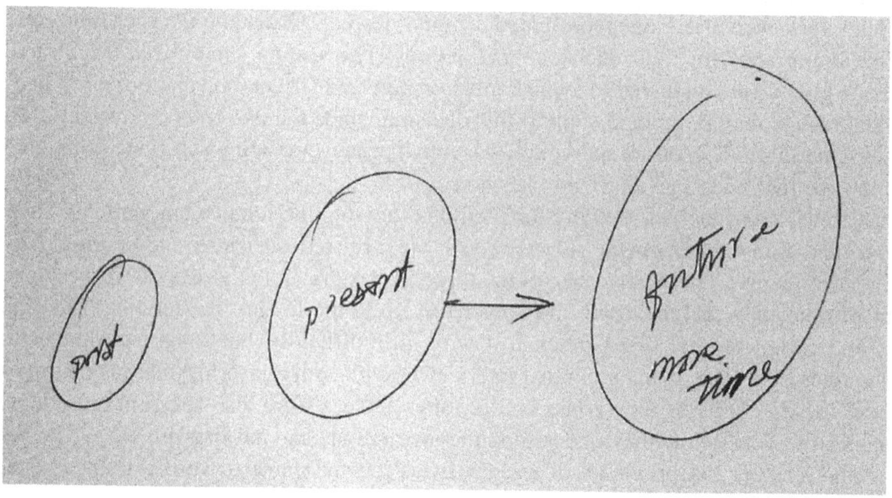

Figure 3.3 A response to the Circles Test for Structure.

Step 3: Measure Temporal Extension

The third step in the Foreseeable Future Multi-measure Method was adapted from Rappaport's Time Line (Rappaport et al., 1985). The original instrument requires interviewees to write significant life experiences in chronological order on a blank strip of adding-machine paper, which they are told represents their entire lifespan. These instructions were adapted so that interviewees are told that the blank strip of paper represents the arrow between the present and the future that they personally drew in the previous step (second Circles Test). The strip of paper can be twisted to represent a cyclic Time Perspective or any other temporal structure. Once the interviewee's understand that the strip of paper represents a timeline they are asked to stick their predefined Motivational Objects (Post-Its) on it in chronological order.

During the empirical test it often became clear that interviewees perceived dependencies between different objects and they were allowed to iteratively relocate the Motivational Objects until they were content with the temporal loci. Next, the interviewee marked units of time on the strip of paper to quantify the Temporal Extension of each object. An example of a completed Time Line is depicted in Figure 3.4. All interviewees understood this step and gave answers. Since most people are concerned with relatively immediate objects, Temporal Extension data is generally skewed to the left. Similarities and differences between groups should thus be analysed using statistical methods that do not assume normal distribution.

Steps 1-3 were sequenced to filter out socially acceptable responses to some degree (criterion 2). One manager insisted that his colleagues, who were to take the interview after him, must also answer the questions in the same order, and said: "Otherwise it's not fair. If I'd known you were measuring the time horizon of my goals then I would have written down different goals." Post-test discussions with managers in the Netherlands, Ghana, and Brazil often involved the interviewee expressing concern about their

Temporal Extension being unacceptably short, but most of them also 'feared' that the results were correct. The Circles Tests and the direct questions may be more prone to socially acceptable responses, though the majority of interviewees appeared to give candid, considered responses.

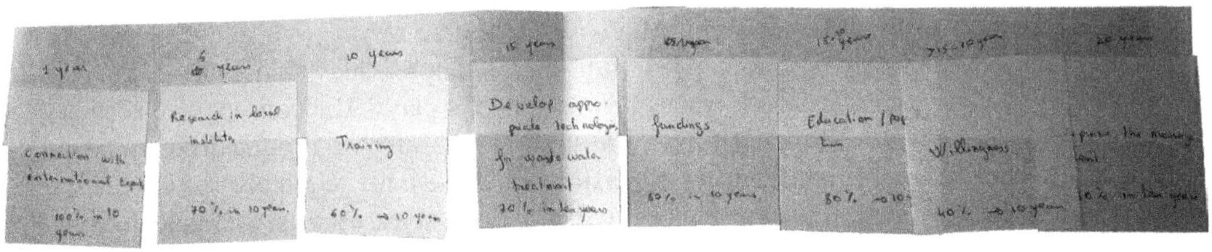

Figure 3.4 Motivational Objects located on a Time Line.

Step 4: Direct Questions

The fourth step in the new method includes 5 statements with blank spaces to be filled in by the interviewee. The interviewees attribute percentage certainties to the terms 'foresee' and 'know', quantify the temporal extent of the foreseeable future, and weight each time zone according to its importance for their work related decisions and actions using percentages. Previous steps are thus cross-examined and augmented with extra information, though each question was treated independently and no reference was made to previous measures.

The first three steps, which are based on visual symbolization using space as a metaphor for time, were consistently understood by interviewees from different cultures during the empirical study (criterion 4). The direct questions used in step 4 were also so straightforward that they appear suitable for cross-cultural studies and useful for checking and augmenting the information produced by the primary measures. Few interviewees required extra help to answer these questions. It is important to note that the direct questions are, however, more prone to socially acceptable answers: The indirect measures should carry more weight in any conclusions.

Step 5: Inventory characterizing Orientation

The final step in the new method involves an inventory with 30 items. The interviewee was asked to specify their personal level of agreement with each statement on a 5-point Likert Scale. Once again, the measures that were adapted for this section cannot be considered equivalent to the originals. Relevant items were selected from two existing inventories. The first 13 items on the new inventory are the complete Future Perspective Scale from the Zimbardo Time Perspective Inventory (ZPTI). As Zimbardo states in his book about this inventory, each of the 6 scales that make up the ZPTI are theoretically unrelated (Zimbardo and Boyd, 2008, Page 52): "Your score on one dimension is unrelated to your score on others." The new method also includes 5 extra items from the ZPTI that concern beliefs about fate. These items were considered the most relevant of the rest. The reason for not including the entire ZTPI is that it is too time-consuming to be incorporated into the new method. The pre-test analysis also revealed some serious issues with using this type of inventory in cross-cultural studies.

Although the ZTPI has been translated into many languages and validated in numerous contexts worldwide the author was confronted with considerable differences in how item content was interpreted by the intended populations, For example, one item reads "I meet my obligations to friends and authorities on time." Firstly, it cannot be assumed that an individual handles obligations to friends and obligations to authorities in the same way. Several interviewees asked whether they should score this item as an average of the two. Secondly, obligations have quite a different nature in different cultures. In Japan, for example, 'giri' is the reflexive 'duty' or 'obligation' which arises from social interaction. But 'giri' never ends: it is a complex and dynamic exchange of favours through a feeling of moral indebtedness. In Brazil and Ghana also different ideas regarding altruism associated with religious beliefs or cultural conventions influenced interviewee's interpretations of 'obligation' as a concept. In the Netherlands this item was generally interpreted more simply as a commitment with a time frame.

A second example of an item from the ZTPI with interpretation issues reads: "Meeting tomorrow's deadlines and doing other necessary work comes before tonight's play." The concept of 'necessary work' for a practical worker in Ghana or Brazil, who may be struggling to survive, is quite different from that for a manager in the Netherlands. And the concept of 'da diversão dessa noite' certainly raised some eyebrows in Brazil. Despite these limitations, the original English phrases were used in Ghana and the validated translations for Dutch, and Portuguese. Brazilian interpreters reworded the Portuguese items so that they would be understood by the Brazilians and Japanese interpreters translated each item from the English original.

In addition to the 18 items from the ZTPI all 12 items from the Consideration of Future Consequences Scale (CFCS) were included, which is designed to measure the degree to which an individual considers the future, rather than immediate, consequences of their decisions (Strathman et al., 1994). These items were translated in the same way as the ZTPI items, which entailed similar interpretation issues. In addition, the CFCS items are lengthier and some interviewees had trouble understanding them. Sum scores were calculated for all three sections of the inventory.

3.3 CONCLUSIONS

In this Chapter the Foreseeable Future Multi-measure Method was described and demonstrated including explanations of the design process. The empirical results incorporated into these descriptions are only meant to communicate how the measures work and to show the nature of the information that is generated. Analysis of these results, to determine the significance of the similarities and differences in the variance of the different variables within and across groups, is dealt with in subsequent Chapters. From a methodological perspective, the main advancement made by this research project is the adaptation and integration of previously distinct measures based on explicit ontological and epistemological premises and a model of Time Perspective. It is hoped that other researchers can also use the Foreseeable Future Multi-measure Method as a basis for future studies of Time Perspectives. The model of Time Perspective may also have useful applications beyond this research project.

By employing multiple measures, the Foreseeable Future Multi-measure Method provides meaningful nuances to the empirical results and facilitates internal checks through triangulation. This is the advantage of a multi-measure method. The main weakness of the method involves the cross-cultural interpretation of items on the adapted inventory (Step 5). This inventory needs to be validated and one may also consider excluding it from future studies because it relies so heavily on construct equivalence and textual translations. The disadvantage of excluding Step 5 is that it would otherwise facilitate comparison of results with those from what is currently the most dominant method used in evaluating Time Perspectives: the Zimbardo Time Perspective Inventory. Additionally, the empirical study did reveal some interesting patterns in the results from the inventory. For example, Managers generally scored higher on the ZTPI Future Perspective Scale than Scientists and Practical Workers in each of the countries. This supports the idea that Managers tend to be biased towards the future, albeit a short-term future. The inventory was eventually included in the definitive interview schedule, for this reason, but results from this measure need to be seen with a particularly critical eye.

To conclude this chapter it is necessary to assess the extent to which the method can produce data, information, and knowledge that forms a defensible basis for ascertaining similarities and differences in the Time Perspectives of people from different cultures and professional backgrounds. The researcher is convinced that the Foreseeable Future Multi-measure Method is fit for this purpose because it generates extensive qualitative and quantitative data that can be cross-checked using the results from different measures and against the backdrop of visiting each interviewee's place of work and living for a short period within the context of each National Culture of interest. The experience of personally interviewing more than three hundred individuals in four countries, including the informal observations and conversations, left impressions about each group that were supported by the results of the structured interviews. The initial scan of these empirical results also revealed that they support the postulation that people from different cultures and professional roles perceive and deal with time in different ways. As stated in the epistemological position, reliability is understood in terms of reflexivity. The validity of the method is demonstrated by rationally explaining the high level of agreement between the experience, the hypotheses, and the empirical results.

Besides the philosophical foundations that imposed requirements on the method, four criteria were defined during the development of the conceptual framework. These criteria guided the development of the method, rather than the initial purpose of aiding the selection of an existing method, so it can be concluded that the Foreseeable Future Multi-measure Method satisfies each of them. A scan of the Motivational Objects provided by interviewees indicated that the procedural steps taken to limit each individual's identity to their Professional Role were successful (criterion 1). Most of the goals and events referred to were work-related. Future studies may focus more on the content of these objects to better understand what motivates people. Criterion 2 prescribes that the method must account for untrue socially acceptable responses. As mentioned earlier, the effectiveness of the ordering of tasks for this purpose is supported by the fact that interviewees wanted to go back and add or remove Motivational Objects, which was prohibited, after realizing that their Temporal Extension would depend on it. Besides the order of tasks, and promising anonymity, more measurable solutions for criterion 2 were not found. On the other hand, the empirical test did confirm that the Foreseeable Future Multi-measure Method facilitates static group comparison over a short period: 309 interviews were held in 20 weeks (criterion 3). The method also incorporates existing measures that had proven useful in different languages and cultures, with a caveat for the inventory step (criterion 4).

The following chapters deal with the next phase of the research project, which was to perform more extensive statistical analysis of the correlations between the variables as a basis for assessing the relevance of the results and for grounding further conclusions and advice.

Chapter 4

An empirical study in Brazil, Ghana, Japan and the Netherlands

Andrew Segrave, Mariëlle van der Zouwen and Wim van Vierssen

This chapter is a slightly adapted version of: Segrave, A.J., Van der Zouwen, M.W., and Van Vierssen, W. 2013. Water Planning: from what Time Perspective? *Technological Forecasting and Social Change.* http:// dx.doi.org/10.1016/j.techfore.2013.08.019. The starting point for that paper is the observation that integrative planning for water technology and infrastructure depends on various social, behavioural and environmental factors. Since water is a common-pool resource, water planning commonly involves addressing the views of diverse groups of people on multiple levels. Inter-group differences, especially if they go unrecognized, can be problematic for planning processes. Research into the Time Perspectives of people fulfilling different roles in the water sector worldwide revealed fascinating differences in how people perceive the foreseeable future. The aim was to characterize the heterogeneity so that it might be explicitly accounted for in planning.

One-on-one interviews were held with 309 managers, practical workers, and scientists in the Netherlands, Ghana, Brazil, and Japan. Each interviewee's Time Perspective was characterized using a multi-measure method (Chapter 3). The most noteworthy conclusions concern differences in the temporal extent of the goals and events that motivate people. Scientists are motivated by objects that lie furthest into the future followed by managers and then practical workers (Group medians: 8.4, 3.5, 1.7 years). Across national cultures, the time horizons of the Japanese and Brazilian interviewees are longest. These conclusions are important because the time horizons considered in planning influence what problems are perceived, what questions are asked, and what solutions are sought.

4.1 INTRODUCTION

4.1.1 Why characterize heterogeneity in Time Perspectives?

In the water sector, integrative planning evolved in response to the perception that fragmented, technocratic practices were preventing sustainable development (Molle, 2008). This paradigm shift echoed developments in environmental planning and planning theories in general from the early 1970s onwards (Mintzberg, 1994; Davoudi, 2001). In theory, integration involves harmonizing the plans of different organisations and sectors on multiple levels, recognizing physical interdependencies (e.g. water-land, quality-quantity) across administrative borders, and heeding the social, economic, and ecological interests of present and future generations. Understandably, some experts argue that complete integration across all of these divisions and scales is impractical (e.g. Biswas, 2004). On the other hand, the problem of fragmentation remains. International leaders have observed that "water security is the gossamer that links together the web of food, energy, climate, economic growth, and human security challenges that the world faces over the next two decades" (Waughray, 2011). Since fresh water is a common-pool resource that connects so many important concerns, it is likely that planners in the water sector will continue to encounter diverse individual and collective interests.

For a mixture of historical, ideological, and practical reasons, it is currently considered proper that the perspectives of public and private stakeholders be included in planning via some form of participation (Webler and Renn, 1995; Campbell and Marshall, 1999). It is generally assumed that the engagement of stakeholders generates more legitimacy, mandate, and support for a plan and that the collective framing of issues is the first step towards designing suitable responses (Barber, 1984; Korthals, 1994; Pahl-Wostl et al., 2008). But participatory approaches have proven to be particularly problematic in practice (Bruns, 2003; Currie-Alder, 2007). Part of the failure of inclusive practices in integrative planning derives from how ambiguity is handled (Brugnach et al., 2012). Problems can be framed, and solutions sought, from various conflicting but equally valid viewpoints (Dewulf et al., 2005). In response, theorists suggest that collective learning be facilitated prior to decision making processes to make use of the diversity in perspectives and knowledge (Pahl-Wostl et al., 2007b; Sterling, 2007; Ker Rault and Jeffrey, 2008; Ker Rault and Jeffrey, 2008; Pearson et al., 2010). Heterogeneity in viewpoints can, however, hamper planning processes; particularly if fundamental differences go unrecognized (Van 't Klooster, 2007; Hulme, 2009; Muro and Jeffrey, 2012). In this Chapter the author explores the idea that one of the most fundamental sources of heterogeneity in perception and behaviour, which is seldom addressed explicitly, may be Time Perspective.

If planning is taken to mean "the design of a desired future and of effective ways of bringing it about" (Ackoff, 1970), then it follows logically that planners implicitly work with basic concepts such as 'the future' and 'time'. There has been criticism of this broad definition (e.g. Mintzberg, 1994), but it suffices here to support the premise that conception of time is necessary for planners. And research has found that time is experienced and conceptualized in fundamentally different ways across individuals and across societies (Boniecki, 1980; Noyes, 1980; Gell, 1992; Levine, 1997). Each individual has a Time Perspective consisting of multiple dimensions that are formed by various factors (Jones, 1994). What makes this important is that the Time Perspectives of planners influence the way they frame problems and conceive solutions: "not only the present state of affairs, but also future and past events in the frame of a subject's Time Perspective co-determine behavior" (Nuttin, 1964). Individuals form groups that have collective views on time, known as Time Frames. The Time Frame of a group can be characterized by the Time Perspectives of the individuals it comprises (Noyes, 1980), but it can also be seen as a cultural characteristic with its own emerging properties (Levine, 1997; Hofstede et al., 2010). The researcher decided to focus on the individual, being the most clearly autonomous entity, and investigate similarities and differences in the Time Perspectives of people fulfilling different roles in the water sector in different national contexts. The aim was to characterize the heterogeneity so that it might be explicitly accounted for in integrative planning.

4.1.2 The Foreseeable Future as a function of National Culture and Professional Role

Variance in Time Perspectives across cultures may be one of the most fundamental sources of heterogeneity in viewpoints of people working within the water sector worldwide as regards planning. Natural hydrogeological resources often extend across administrative and national borders. This means that people from different cultures have shared interests in a common resource (Enserink et al., 2007). As a result, high level policies and laws, such as the EU Water Framework Directive, focus on the scale of river basins and stipulate international cooperation (WFD 2000/60/EC). Response strategies for the main long-term challenges facing the water sector at the moment, such as Climate Change, also require international cooperation (Hulme, 2009). Various studies have compared Time Perspectives across countries and revealed significant differences (Levine, 1997; Raju, 2003; Hofstede et al., 2010). Although culture is defined a shared quality that is intrinsic to the 'self' rather than being a framework or setting, and regional cultural differences and dynamics are recognized (Geertz, 1973; Heine et al., 2002), national borders were accepted as rough divisions between cultures as have previous studies (e.g. Enserink et al., 2007).

In addition to culture, variance in Time Perspectives across professional roles is likely to be one of the most basic sources of the heterogeneity in the viewpoints that planners are confronted with in the water sector (Whitney and Smith, 1983; Lichtenstein et al., 1997; Wageman et al., 2012a). People with different personality characteristics and competencies are attracted to different roles. And to attain these different positions they also follow different education and career pathways. While working in a certain role, by a process of 'identity negotiation', individuals also establish mutual expectations of one another to form groups (Goffman, 1959; Swann, 2009). So people fulfilling different roles in the water sector are likely to have quite different viewpoints. One of the main challenges facing planners is to translate long-term, abstract visions and policies into strategies and concrete actions. This translation depends on cooperation across roles; between people who work with abstract concepts and theories and people who perform practical tasks. Research has found this to be a problematic process (Burby, 2003; Brody and Highfield, 2005; Cheng and Daniels, 2005; Agrawal and Lemos, 2007) and this is also the experience of the author.

The independent variables were thus defined as National Culture and Professional Role. Based on earlier research (Levine, 1997; Hofstede et al., 2010), four countries were chosen to represent the variance between National Cultures: the Netherlands, Ghana, Japan, and Brazil. Since public and private water sectors are likely to differ considerably, to assess the influence of the independent variables with a sample of just four countries it was necessary to control this variable. Only organisations from public sectors were included in the present study but future research may include this variable to, for example, assess possible implications of privatization. As regards Professional Roles, three sample groups were defined to represent the level of abstraction from theory to practice:

Scientist: a specialist in one of the physical or natural sciences related to water management.
Manager: someone who controls resources, expenditures, and the direction of (part of) an institution, which in turn directly influences water resources.
Practical Worker: someone who physically intervenes in the management of water resources, generating material outcomes with their individual actions.

These roles were chosen for practical reasons; considering the structures of the water sectors in the countries of interest. The option of including policy makers was considered but the 3 groups that were selected are assumed to represent variation across different Professional Roles sufficiently and organizing interviews with additional groups was likely to be impractical. Two main research questions were posed to direct the analysis:

- What are the similarities and differences in the Time Perspectives of people with different Professional Roles within a given National Culture?
- What are the similarities and differences between National Cultures in the Time Perspectives of people with comparable Professional Roles?

Various methods exist for characterizing how people conceptualize and deal with time. Zimbardo and Boyd (2008) claim that no two individual's attitudes toward time are identical. Previous empirical research also showed that there is limited overlap between the various markers of Time Perspective and that the outcomes tend to depend on the method used (De Volder and Lens, 1982; Gjesme, 1983; Rappaport et al., 1985). Adams and Nettle (2009) found significantly correlated markers of Time Perspective, but the strength of correlations was not strong. Methodologies also vary with different philosophies of time, though the ontological and epistemological grounds for existing methods are generally unclear. For these reasons, the author developed a new multi-measure method that integrates existing measures within a model of Time Perspective (Chapter 3). This method rests on the idea that the dimensions of Time Perspective that are likely to have the greatest influence on integrative planning concern forethought. Previous research has argued that motivation for decisions and behaviour is related to the present anticipation of future goals (Polak, 1973; Simons et al., 2004). More specifically, Nuttin (1964) defined the future as "the time quality of the goal object; the future is our primary motivational space." De Volder and Lens (1982) used this premise to integrate the dimensions of Time Perspective that concern the future into cognitive theories of human motivation. Building on this logic, the dependent variable is characterized using two dimensions:

- Motivational Objects: causally efficacious goals and events defined by interviewees
- Extension: the span of time into the future that is populated by Motivational Objects

Motivational Objects must influence decisions and actions to be considered real (causally efficacious). There has been much criticism of methods that lump Motivational Objects with ineffectual wishful thinking (Klineberg, 1967; Wohlford and Herrera, 1970; Lessing, 1972). But people have fundamentally different definitions of what is 'foreseeable': To what degree can a belief be justified as true before the object of the belief exists? For this reason a third dimension is added to the dependent variable:

- Certainty: A percentage likelihood attributed to Motivational Objects by the interviewee

The dependent variable, called 'Foreseeable Future', was thus defined as the temporal Extension of an individual's Motivational Objects and the Certainty attributed to these objects. Since Extension is a measure that is relative to the structure of time, which is the subjective concept of time itself including definition of the time zones, it was necessary to characterize this aspect as well. Likewise, if an individual is biased towards or against the future compared with the past and the present (Orientation) then this may influence the Extension of their Motivational Objects. Foreseeable Future is thus characterized relative to the Structure and the Orientation of each individual's Time Perspective. Figure 4.1 is a simplified visualization of the relationships between these variables. The independent variables (National Culture and Professional Role) are seen to influence an individual's entire Time Perspective. Of the multiple dimensions that might be used to characterize a Time Perspective, the researcher is specifically interested in a subset that is used to define the 'Foreseeable Future'. Given that these specific dimensions can only be interpreted relative to the more fundamental dimensions of 'Structure' and 'Orientation', however, it is necessary to characterize the latter two as well. An overview of the measures used to characterize each of these dimensions is given in the next section.

4.2 MATERIAL AND METHODS

The sample population was surveyed using structured one-on-one interviews as described in Chapter 3. This allowed the researcher to sense first-hand whether or not the interviewees understood the abstract subject matter but limited the sample size. A standard letter was written in English, translated into the local languages, and sent to the main water organisations in each country. This letter requested interviews with Practical Workers, Managers and Scientists and provided a definition of each role. The final sample

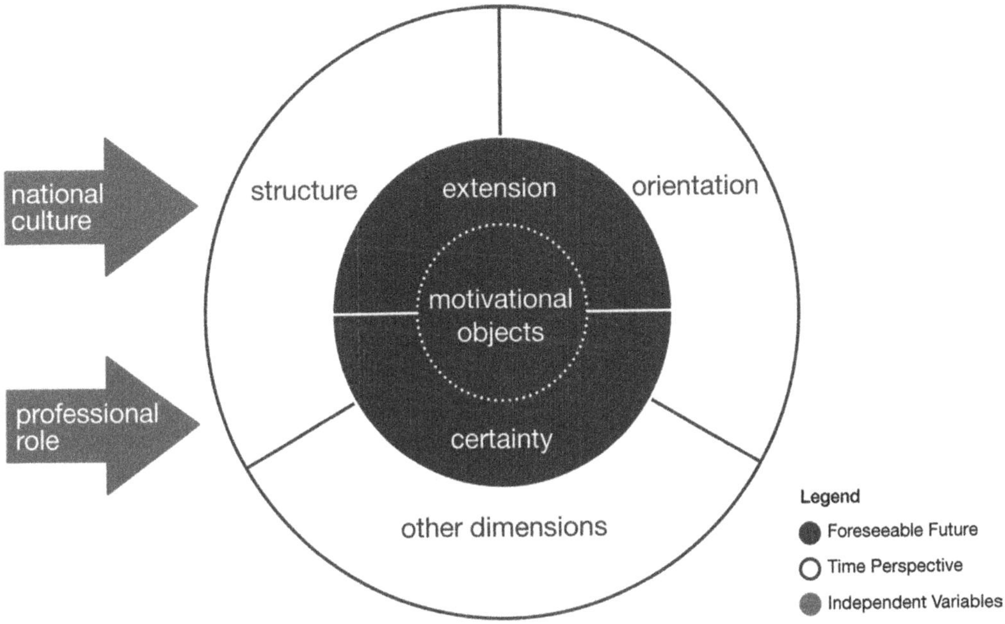

Figure 4.1 Visualization of relationships between variables.

population included 309 individuals who were categorized according to National Culture and Professional Role into 12 groups (Table 4.1). No statistically founded sampling method was used, since the researcher depended on local contacts to arrange the interviews.

A ring binder containing one punched plastic pocket per interviewee was prepared for each country. Every plastic pocket included:

- 1 cover page with anonymous interviewee background information
- 1 blank piece of A4 paper
- 1 60cm long, 5 cm wide, blank strip of adding-machine paper
- 1 printed questionnaire with 5 direct questions
- 1 printed inventory with 30 items and a 5 point Likert scale

The researcher also brought certain standard materials to each interview:

- 1 interview schedule printed in English
- 1 interview schedule printed in the interpreter's language
- Yellow post-it notepads
- 10 questions, each printed on a separate card

The interview schedule prescribes every step and every cue in method explicitly. Besides dictating the order of tasks and the instructions this schedule allowed control over the translation and interpretation of concepts, which is essential for cross-cultural comparisons (e.g. Hambleton et al., 2005). It should be noted that existing measures that were adapted and integrated into the Foreseeable Future Multi-measure Method (Chapter 3) cannot be considered equivalent to the originals. The pre-test considerations concerning translation of the methods for each language and culture are described in section 3.2.2. What follows is a brief overview of the procedure followed in each interview.

Table 4.1 Number of individuals per group.

National Culture	Practical worker	Manager	Scientist	PRs Lumped
	Professional Role			
Netherlands	22	21	21	64
Brazil	18	39	21	78
Japan	29	23	36	88
Ghana	27	26	26	79
NCs Lumped	96	109	104	309

A single interviewer conducted all of the interviews in person together with a local interpreter in each country. The structured interview itself is divided into 5 successive steps that were sequenced to account for socially acceptable responses and to facilitate triangulation. These steps each focus on a specific aspect (Section 2.3.6), summarized as follows:

Step 1: Define Motivational Objects

The interviewee wrote down their most important goals and future events, each on a separate post-it sticker. Only those goals and events that had influenced their decisions or actions qualified as Motivational Objects. The interviewee attributed a percentage Certainty to each object to indicate the probability of it occurring in the future.

Step 2: Characterize Structure and Orientation

The interviewee symbolized their concept of time using two adapted versions of Cottle's Circles Test (Cottle, 1967). One circles test represented the interviewee's Orientation to each time zone as regards its importance for their decisions and actions. The second symbolized the Structure of time itself, including an arrow showing the path of time. A textual legend was included to explain each illustration.

Step 3: Measure Temporal Extension

The interviewee stuck the post-it stickers from Step 1 in chronological order on a blank strip of adding-machine paper, which represented the arrow showing the path of time that was drawn in Step 2.This is an adapted version of Rappaport's Time Line (Rappaport et al., 1985). A temporal scale was then written on the timeline to indicate the Extension of the interviewee's Motivational Objects.

Step 4: Direct Questions

Five direct questions by completion of sentences were used to check and supplement the results from the indirect methods used in steps 1 – 3.

Step 5: Inventory characterizing Orientation

The interviewee specified their level of agreement on a 5-point Likert Scale for 30 items, including the Future Perspective items from the Zimbardo Time Perspective Inventory and the entire Consideration of Future Consequences Scale.

The five steps generally took 30 to 60 minutes to complete. If an interviewee obviously didn't understand a question then the interviewer explained the meaning and took note of everything that was said. This was also true for any other digressions from the schedule or interesting comments that were made. The interviewee drew or wrote down their responses to all of the questions and this, together with the interviewer's notes, made up the record of the interview. All 309 interviews were digitized and the numeric data was tabulated for input into statistical analysis software. Interviewees also characterized themselves according to the independent variables (National Culture and Professional Role) and standard confounders (Age, Gender etc.) on a background information form.

4.3 RESULTS AND DISCUSSION

The results and discussion are combined in this section to enrich the quantitative records of the interviews with qualitative information gained through the procedure and to enable the significance of the empirical data to be explored efficiently. Foreseeable Future is relative to Structure and Orientation (Figure 4.1), so to start with these more fundamental aspects are analysed across National Cultures and Professional Roles. The similarities across groups as regards Structure and Orientation are more striking than the differences, so the next step is to make comparisons in terms of the Foreseeable Future: first across National Cultures and then across Professional Roles. Conclusions in the next section are based on these findings. Each interviewee provided additional data on their Age, Gender, Income, Health, Education, and Religion. Some correlations were found between these and other variables however they were either encapsulated within the independent variables, such as Income being an aspect of Professional Role, or not significant or influential enough to be considered confounders.

4.3.1 Structure

The results of the Circles Test for Structure are recorded in Table 4.2. The schematic drawings in each cell represent the mode for the corresponding group, which resulted from digital analysis of the codified

Table 4.2 Mode per group for the Circles Test for Structure.

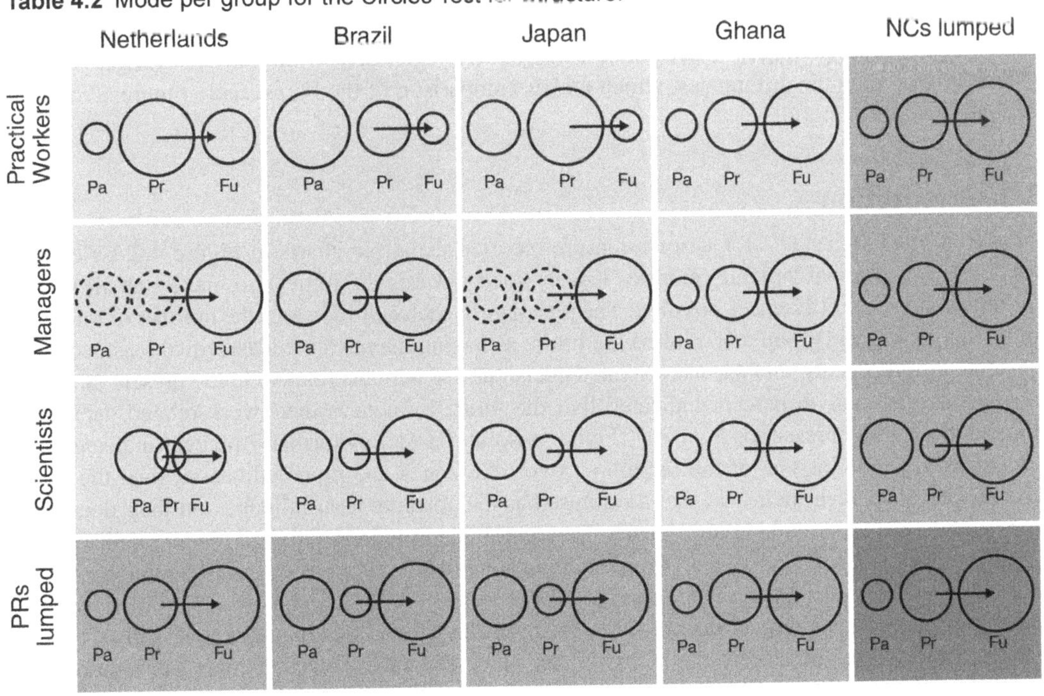

original illustrations. The relative size of the three circles in these drawings indicates the amount of time in each time zone: the bigger the circle the longer the duration. The arrangement of the circles, as regards order and overlap for example, illustrates how interviewees picture the relationships between the time zones. And the arrows signify the path that time is perceived to take, specifically between the future and the present. Textual legends were used to verify the meaning of these visual attributes. Since these drawings of Structure are characterized by several interrelated aspects, including the degree of overlap, the arrangement and relative size of the circles, and the directionality and (non) linearity of the timeline, similarities and differences are not calculated using statistical methods. Bimodal results (e.g. managers in the Netherlands and Japan) are recorded using dashed lines to show both modes.

Most interviewees (80 percent) drew the past, present, and future following a linear path from left to right in that order. A further 11 percent drew the circles following a linear path but with the time zones in a different order. Just 9 percent drew the circles in a cluster; indicating a nonlinear path of time. Of the 9 percent who drew a nonlinear timelines, half drew a cyclic path and half drew a zigzag back and forth between the time zones. The mode for all groups, except scientists in the Netherlands, was to draw each time zone separately; without any overlap (79 percent of interviewees). Of the 65 individuals who did draw overlapping time zones, 40 percent were Dutch and 21 percent were specifically Dutch Scientists. Less than 1 percent of the Ghanaian interviewees drew overlapping time zones. Since the interview questions did not mention overlap, the differences between National Cultures in this regard are likely to represent a subtle difference in how the time zones are conceptualized.

Most individuals drew the circle for the future as the biggest and that for the past as the smallest; indicating that the future would have longer duration than the present and that the past comprises the least amount of time. Notable exceptions include Scientists, who generally indicated that the present comprises the least amount of time. Comparison across National Cultures reveals two main modes: Japanese and Brazilian interviewees indicated that the present has the shortest duration whereas the Ghanaian and Dutch interviewees saw the past as shortest. There was no clear mode for Managers in the Netherlands and in Japan, with equal numbers indicating that the past is longer than the present as the opposite. In all cases the meaning of the depictions was confirmed using textual legends. The similarities as regards Structure are more striking than the differences, which makes comparison of the Foreseeable Future across groups easier and more meaningful.

4.3.2 Orientation

Results from the Circles Test for Orientation are reported using pie charts in Figure 4.2. Orientation is translated into an ordinal variable (more or less inclined towards the future), to make statistical analysis of the similarities and differences possible. The pie chart in each cell indicates the number of interviewees from the corresponding group who ranked the future as having the absolute least, equal least, equal most, or absolute most influence on their work related decisions and actions, relative to the present and the past.

Of all interviewees, 46 percent indicated that the future influences their work related decisions and actions more than the present or the past ('Future absolute most'). A further 7 percent indicated that the future is as influential as one of the remaining two time zones and more influential than the other. Of the remainder, 17 percent indicated that the future has the absolute least influence on their decisions and actions with 'equal least' making up a further 7 percent. In the category 'Other' 9 percent indicated that all time zones are equally influential and 14 percent indicated that the future is more influential that one of the time zones but less influential than the other.

There appears to be a general bias towards the future in Brazil and Ghana, which is also true for the Managers in each country except the Netherlands. Since the dataset doesn't meet the requirements of parametric tests, non-parametric techniques are used to compare groups. If Professional Roles are lumped, a Kruskal–Wallis test indicates that the distributions of the Circles Test for Orientation are not equivalent across categories of National Culture: H(3)=24.36, .<.05. Taking the Dutch as a reference category, 3 post hoc Mann–Whitney U tests reveal that the differences are only statistically significant when the Dutch are compared to the Japanese (U = 1912.50, . < .017, r = -.27). If National Cultures are lumped, a Kruskal–Wallis test shows that the distributions do not differ significantly across categories of Professional Role (.>.05). Differences across Professional Roles within each National Culture are also not statistically significant. The experience of conducting the interviews left the impression that the Brazilians and the Ghanaians were most focused on 'the future' whereas the Japanese and the Dutch appreciated the importance of all three time zones.

To cross check the Circles Test for Orientation, interviewees were given a direct question at the end of the interview asking them to provide a numeric ratio between the past, present, and future to estimate relative levels of influence on their work related decisions and actions (e.g. Past: 20; Present: 30; Future:

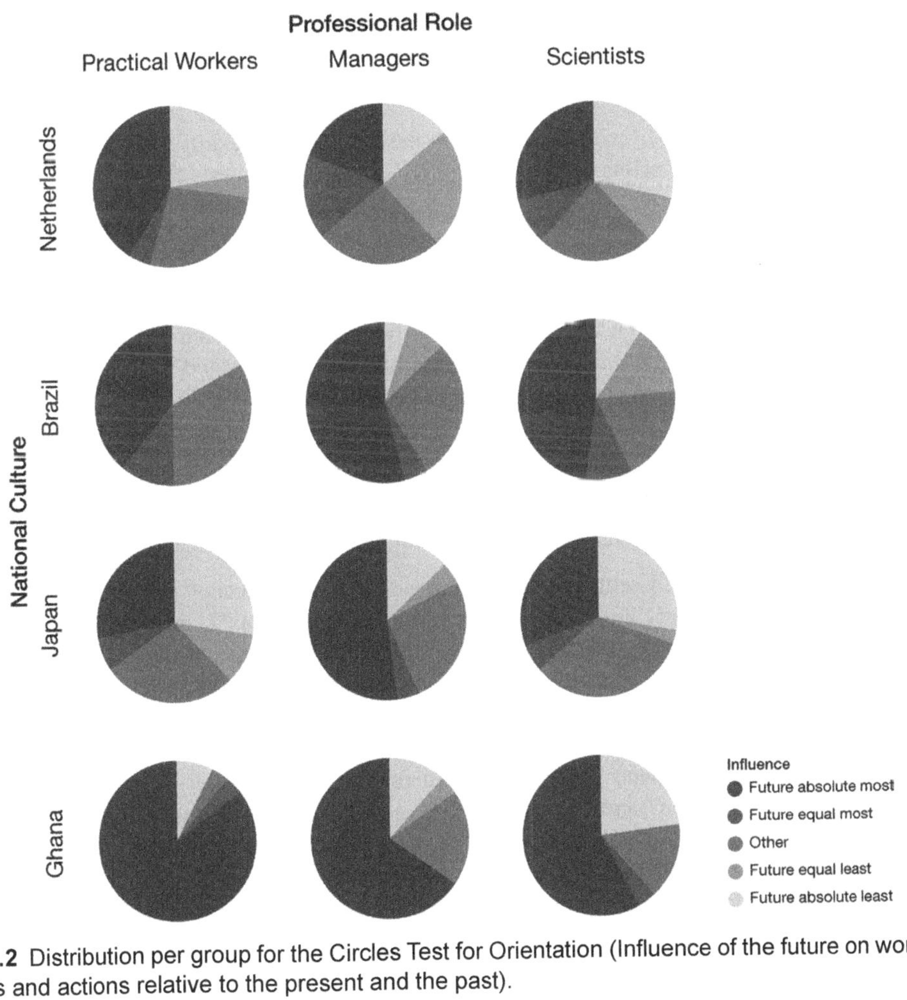

Figure 4.2 Distribution per group for the Circles Test for Orientation (Influence of the future on work related decisions and actions relative to the present and the past).

40). In general, interviewees who drew the circle representing the future as the biggest (having the most influence) also indicated that their work related decisions are primarily based on expectations about the future. This correlation is moderate (. = .39) and significant at the .01 level (2-tailed). The correlation between these measures is particularly strong within the group Scientists, but for Mangers it was weaker. It was also moderate to strong in all National Cultures except Brazil. The direct question thus supports the Circles Test results by methods triangulation.

As a second cross check interviewees completed an inventory with 30 items from prominent instruments including the Consideration of Future Consequences Scale (CFCS) and subscales from the Zimbardo Time Perspective Inventory (ZTPI). Comments made during the interviews, however, raised concerns about construct equivalence across cultures. For example, one ZTPI item reads "I meet my obligations to friends and authorities on time." The concept 'obligation' can have quite different connotations in different cultures. In Japan, for example, 'giri' is the reflexive 'duty' or 'obligation' which arises from social interaction. But 'giri' never ends: it is a complex and dynamic exchange of favors through a feeling of moral indebtedness (Benedict, 1989). Brazilian interviewees also mentioned perceiving considerable differences between obligations to friends and obligations to authorities. Furthermore, interviewees had difficulty understanding the CFCS items. Another recent study also found this to be the case (Mckay et al., 2012). While recognizing the limitations of these methods the researcher found a weak correlation (. = .11) significant at the .05 level (2-tailed) between the ZTPI scores and Orientation towards the future as characterized by the Circles Test. CFCS scores are also significantly correlated at the .01 level (2-tailed) with the Circles Test results, however weakly (. = .24). These correlations confirm the primary results.

4.3.3 Foreseeable Future

The Foreseeable Future is characterized by the Extension and Certainty attributed to Motivational Objects. The results for these three dimensions are summarized in Table 4.3 using medians because the distributions are skewed towards fewer Motivational Objects (Count), higher Certainty, and shorter Extension.

The word 'foresee' is quite ambiguous, although it is a common term associated with integrative planning in the water sector. To evaluate this term, interviewees were presented with 2 statements with blank spaces:

1. As regards my work, when I "Foresee" something then I am _% certain that it will happen.
2. As regards my work, when I "Know" that something will happen then I am _% certain.

The median certainty associated with the word "Foresee" was 68 percent and that for "Know" was 86 percent. The median Certainty attributed to Motivational Objects was 71 percent, which is 3 percent more than that associated with "Foresee" and 15 percent less than that for "Know". The results for 11 of the 12 groups showed that the Certainty attributed to Motivational Objects is slightly higher than that for "Foresee" and significantly less than that for "Know". As regards certainty, the Motivational Objects used to define the Foreseeable Future can thus be said to match the interviewees' definition of "Foresee". The notable exception was Japan. Japanese interviewees attributed 77 percent certainty to their Motivational Objects, 60 percent to "Foresee" and 78 percent to "Know". This result may derive from uncertainty avoidance (Hofstede et al., 2010). On the other hand, the translation of questions into Japanese proved more difficult than for other languages and the word "Foresee" is quite ambiguous. A native English speaker may also attribute different levels of certainty to synonyms of "foresee", such as "predict" or "anticipate".

Direct questions were used at the end of each interview to cross examine the Extension results. Interviewees were asked to estimate how far they can foresee into the future as regards important affairs in their field of work. This self-reported time horizon was generally shorter than the Extension of interviewees' Motivational Objects, which may be due to the wording of the questions, aversion to appearing overconfident, and numerous other factors. Even so, there is a moderate correlation (. = .28) between the self-reported extent of the foreseeable future and the Extension of Motivational Objects, which is significant at the .01 level (2-tailed). The Extension results are thus supported by methods triangulation. The results associated with Motivational Objects (Table 4.3) are considered more trustworthy than the results of the direct questions, because the primary method filters out socially acceptable answers and uses actual goals and events.

Table 4.3 Attributes Median Netherlands of Motivational Objects.

	Median	Netherlands	Brazil	Japan	Ghana	NCs lumped
Practical Worker	Count	8.0	6.0	4.0	6.0	6.0
	Certainty (%)	68.4	73.0	80.0	64.3	71.4
	Extension (years)	2.2	1.8	3.0	1.3	1.7
Manager	Count	11.0	8.0	5.0	6.0	7.0
	Certainty (%)	69.3	75.0	82.5	70.0	73.3
	Extension (years)	1.9	3.9	5.8	2.2	3.5
Scientist	Count	9.0	8.0	6.0	7.0	7.0
	Certainty (%)	67.3	71.1	72.6	67.3	68.2
	Extension (years)	4.0	10.0	18.5	4.7	8.4
PRs lumped	Count	9.0	8.0	5.0	6.0	7.0
	Certainty (%)	67.6	72.4	76.8	68.0	71.1
	Extension (years)	2.8	4.9	6.4	2.1	3.7

The results for Extension are particularly suited to graphical representation because space is commonly used as a metaphor for duration. This data is also especially relevant for planning; for example when it comes to phasing and selecting time horizons. For these reasons the Extension data is presented in detail using box plots to illustrate the distributions for each subgroup (Figure 4.3). The box represents the 25th to 75th percentile. The line in the middle represents the 50th percentile. And the whiskers represent the highest and lowest values that are not outliers (1.5 to 3 times the interquartile range: circles) or extreme values (>3 times the interquartile range: stars).

Discussion of these results is organized into a comparison first across National Cultures and then across Professional Roles. The datasets do not meet the requirements for parametric tests. For this reason, the statistical significance of the similarities and differences across groups was analysed using the Mann–Whitney U test and the Kruskal–Wallis one-way analysis of variance by ranks, which have a null hypothesis that the populations have equivalent distributions. A Bonferroni correction was applied to calculate the critical value for significance in 3 ($\rho < .017$) and 6 ($\rho < .0083$) post hoc tests. The group sizes were not large, but greater than 5 in all cases (Table 4.1). Besides analysing similarities and differences, the author also investigated possible correlations. Spearman's Correlation Coefficient (ρ) was used because the data is skewed with outliers and because several variables, such as Orientation, are ordinal. In psychological research, Cohen's (1988) conventions are used to interpret effect size. A correlation coefficient of .10 represents a weak association; .30 indicates a moderate correlation; and .50 or larger is a strong correlation.

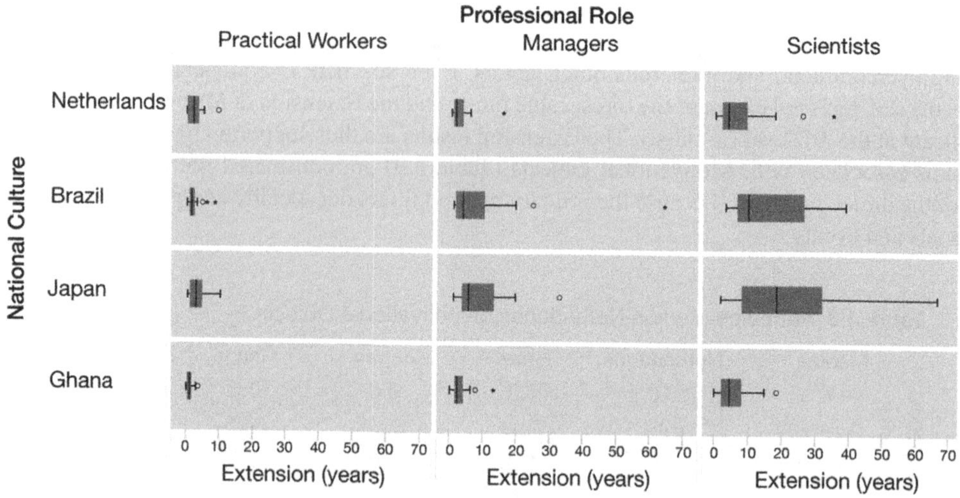

Figure 4.3 Box- plots comparing the temporal extent of motivational objects across interviewees grouped by National Culture and Professional Role.

4.3.4 Comparison across National Cultures

Comparison of Extension across National Cultures reveals that Japanese interviewees are motivated by the most temporally distant objects (Mdn: 6.4 yrs.), followed by the Brazilians (Mdn: 4.9 yrs.), the Dutch (Mdn: 2.8 yrs.), and the Ghanaians (Mdn: 2.1 yrs.). If Professional Roles are lumped, a Kruskal–Wallis test shows that the distributions for Extension differ significantly across categories of National Culture: $H(3)=47.22$, $\rho <.05$. Mann–Whitney U tests indicate that these differences are significant across 4 of the 6 subcategories ($\rho < .0083$). Interviewees from Japan and Brazil are motivated by objects that lie further into the future than those of interviewees from Ghana and the Netherlands. The effect size for the differences is moderate. The differences between Japan and Brazil are not significant, nor are those between Ghana and the Netherlands.

When Professional Roles are differentiated, Extension is significantly different ($\rho <.0083$) across Dutch and Japanese Managers and across Dutch and Japanese Scientists. The Practical Workers in these two countries do not differ significantly. The differences across Dutch and Brazilian groups are not significant within any specific Professional Role. In contrast, the significance of differences across Ghanaian and Japanese interviewees holds for each Professional Role. In the Ghanaian and Brazilian comparisons this is only true for the Scientists. For Managers and Scientists in the comparisons across Brazil and the Netherlands and across Brazil and Ghana the null hypothesis is retained, in part, due to the restrictive critical value for significance ($\rho <.0083$).

The experience of conducting the interviews left the impression that the Brazilian Managers and Scientists were actually motivated by goals and events on considerably longer time horizons than their Dutch or Ghanaian counterparts. This may originate in part from perceived necessity; there remains much to be done in the Brazilian water sector that is unlikely be achieved within the next decennia and these developments can literally save lives. The Brazilians also seemed to have a very strong sense of optimism, with almost utopian views compared with the pragmatic perspectives of the Dutch and Ghanaian interviewees.

Dutch water supply companies have been working on capacity building projects with Ghana Water Company Ltd. since 2006, which may explain some of the similarities in Foreseeable Future. Even so, the relatively short time perspectives of interviewees in Ghana and the Netherlands are likely to have evolved for quite different reasons. In the Dutch water sector there is a sense that the fundamental purpose or mission has been achieved and a focus on pragmatic management, maintenance, and gradual optimization. The vital infrastructure has been working effectively for decades, which can't be said of Ghana. In Ghana there seemed to be a feeling of dependence on greater powers and especially on politics. Ghanaian interviewees conveyed the sentiment that little can be planned beyond the next elections due to reliance on the associated (development) financing.

Interviewees in Ghana, Brazil, and The Netherlands seemed accustomed to having personal goals and priorities whereas Japanese people tended to presume that such matters would be collectively defined by a group. This was one of the main pre-test translation issues. Japanese interviewees used the concept 'mokuhyo', which refers to various concepts ranging from an 'aim' to what might best be described as a 'guiding principle'. One problem with this concept was that a 'mokuhyo' is often seen as timeless; never to be realized. An alternative translation was found in the word 'kadai', which is generally associated with a certain time frame and translates as 'task' in English. But a 'kadai' is not necessarily related to an individual's view and may be obligatory. Conversations about these constructs with Japanese interpreters supported previous research showing that the concept of motivation is likely to differ fundamentally between individualistic and collective societies (e.g. Hofstede et al., 2010). The Tohoku earthquake and tsunami on 11 March 2011 may also have influenced the Time Perspectives of Japanese interviewees. The Japanese were motivated by fewer, abstract principles on a longer time horizon.

Kruskal–Wallis tests indicated that the mean number of Motivational Objects reported by interviewees (Count) differed significantly across National Cultures when Professional Roles are lumped (.<.05). This is also true for the Certainty that interviewees attributed to their Motivational Objects. Taking the Dutch group as a reference category, Mann–Whitney U tests showed that the differences across National Cultures remain significant when each culture is compared separately (. < .017). The differences across National Cultures concerning Count also remain significant within each category of Professional Role. Overall, the Dutch wrote down the most objects (Mdn = 9) and Japanese people reported the least (Mdn = 5). Differences in the Certainty attributed to Motivational Objects across categories of National Culture are due to the relatively high certainties given by Japanese interviewees in all subgroups (Mdn: 77%). Other National Cultures did not differ significantly (.>.0083). It seemed socially acceptable to have numerous goals in the Netherlands, whereas in Ghana people may see this as being a little immodest or unrealistic. The fact that interviewees in the Netherlands and Brazil reported more Motivational Objects than those in Japan and Ghana may also indicate that interviewees in these countries are motivated by a wider range of issues or that they differentiate between objects at a more detailed level. It may be interesting to investigate the contents of the Motivational Objects in future research.

4.3.5 Comparison across Professional Roles

If National Cultures are lumped and Extension is compared across interviewees grouped by Professional Role then it appears that Scientists (Mdn: 8.4 yrs.) are motivated by objects on a more distant time horizon than Managers (Mdn: 3.5 yrs.), and that Practical Workers have the shortest Extension (Mdn: 1.7 yrs.). This pattern is also evident within each National Culture (Figure 4.3). If Professional Roles are given an ordinal rank according to the level of abstraction from theory to practice a strong correlation (. = .53) is found between Professional Role and Extension, which is significant at the .01 level (2-tailed). People working in more abstract roles are motivated by objects that lie further into the future. Kruskal–Wallis tests show that Extension differs significantly across categories of Professional Role when National Cultures are lumped: H(2)= 83.77, .<.05. Mann–Whitney U tests reveal that these differences are significant across all subcategories moderate to strong effect sizes. Extension is significantly different (.<.017) across categories of Professional Role within each National Culture and in all cases except between Practical Workers and Managers in The Netherlands (U = 198.50, . = .43). This similarity may be due to Dutch organisations being less hierarchical than those in other countries. Dutch Managers might also grow into their role via practical experience. The focus on maintenance in the Dutch water sector may also shorten Managers' Extension to that of Practical workers, since they are not striving to achieve grand plans like the Managers in Brazil.

Kruskal–Wallis tests indicated that the mean number of Motivational Objects reported by interviewees (Count) differed significantly across Professional Roles when National Cultures are lumped (.<.05.). Practical Workers had significantly less Motivational Objects (Count) than Managers and Scientists (. < .017) but Managers and Scientists did not differ significantly. Practical Workers may have reported less Motivational Objects because they are responsible for a narrower range of matters or because they are less accustomed to thinking about and writing down their goals. Practical Workers did generally struggle completing this step, whereas Managers had typically listed their most important future goals and events within a few minutes. Comparison across Professional Roles within the Netherlands revealed the same pattern, but in Brazil only the Practical Workers and Managers reported significantly different numbers of Motivational Objects and the variance across Professional Roles was not significant within Japan or Ghana.

Differences in the Certainty attributed to Motivational Objects across categories of Professional Role are due to the relatively high certainties given by Managers in Japan (Mdn: 83%) and Brazil (Mdn: 75%). Managers attributed the most certainty to their Motivational Objects (Mdn = 73%) and, in general, Scientists were least certain that their objects would materialize (Mdn = 68%). Certainty did not differ significantly between Scientists and Practical Workers, and only in Japan did Certainty differ significantly across all Professional Roles. The researcher investigated the significance of correlations between Extension and the number of Motivational Objects reported, and between Extension and the Certainty attributed to these goals and events, but none were significant. Even so, the fact that Scientists were less certain that their objects would eventuate is logically related to these objects being further into the future. And perhaps Managers need a greater level of certainty to be able to make decisions or act; reducing their time horizon to a practical extent. In the conclusions the relationships between the results of the different measures are described in more detail.

4.4 CONCLUSIONS

One-on-one interviews were held with 309 people working as Managers, Practical Workers, or Scientists in the Netherlands, Ghana, Brazil, or Japan. Each interviewee's Foreseeable Future was characterized using a multi-measure method. Comparison of how interviewees visualize the Structure of time revealed more similarities than differences across all groups. The most noteworthy difference was that Scientists and interviewees from Japan and Brazil drew the present as having the shortest duration whereas the mode for all other groups was to indicate the past as being shortest relative to the present and future. Dutch Scientists also drew overlapping time zones, indicating interdependence, which was unique. There were no remarkable differences across groups as regards Orientation to the time zones. The results of the Circles Tests that were used to characterize these dimensions were confirmed by methods triangulation using direct questions and an inventory. Since the Foreseeable Future is relative to Structure and Orientation, the similarities form a consistent basis for comparison across groups.

Foreseeable Future was characterized by determining the temporal Extension of each individual's Motivational Objects and the Certainty that they attributed to these objects. Scientists were motivated by objects on a more distant time horizon than Managers and Practical Workers had the shortest Extension.

These differences were significant within all National Cultures and across all Professional Roles except between Practical Workers and Managers in The Netherlands. As regards differences across cultures, the Extension of Japanese and Brazilian interviewees was significantly longer than that of the Dutch and the Ghanaians. Direct questions confirmed these results by methods triangulation. Future research might investigate possible interactions between National Culture and Professional Role in determining Extension. There was no correlation between Extension and the number of Motivational Objects reported, or between Extension and the Certainty attributed to these goals and events. Even so, the fact that Scientists were generally less certain that their objects would eventuate is logically related to these objects being further into the future. It is interesting to note that the groups with the longest Extension across Professional Roles (Scientists) and across National Cultures (Japan and Brazil) also indicated that the present has the shortest duration. It seems logical that the present be perceived as comparatively shorter by people with longer Extension into the future. And it is fascinating to find that these groups also appear to perceive the past as comprising more time than the present, whereas for other groups the reverse was true. Later research may investigate this possible relationship further. Various potential confounders were considered but none were found to be problematic.

A quantified Foreseeable Future may be useful for environmental law in cases related to liability and, for example, the Precautionary Principle. Economic models that incorporate Temporal Discounting could also make use of a quantified Foreseeable Future. Subsequent research may attempt to extrapolate the Foreseeable Future for countries other than the four in the sample. It may also be interesting to investigate whether the Foreseeable Futures of the groups interviewed have had perceivable consequences for the way the different water sectors handle investments in technology and infrastructure, both concerning the operationalization of plans and scope of their science systems. When it comes to motivation for behaviour, the future is the time quality of the goal object (Nuttin, 1964). The Time Perspectives of different groups in the water sector must, therefore, influence the way they frame problems and conceive solutions. Differences in the Foreseeable Future as seen by the various actors involved in integrative planning for investments in water technology and infrastructure are likely to increase the ambiguity of the decision making processes and seriously influence the outcomes. For example, the Foreseeable Future of Brazilians could enable them to leapfrog the Dutch water sector during the coming decennia while that of the Ghanaians may impede such long-term development. Similarly, concerning countries with highly developed water sectors, the Dutch may be at risk of succumbing to the handicap of a head start while the Japanese use their long-term perspective to keep ahead.

Insight into the Foreseeable Future of different groups could enable planners to phase projects using time periods that keep the different groups motivated. Likewise, futures researchers might use these insights to determine time horizons for their studies and to translate their findings into indicators that mangers are more likely to act on. Wicked problems in the Water-Energy-Food-Climate nexus require that global long-term thinking be translated into local short-term action. Insight into the Foreseeable Future of different groups may assist planners in bridging this gap. Scientists have the longest Extension, the highest in-group variance, and are also typically accustomed to international collaboration. Besides the different roles of individual experts as described by Pielke (2007), the science sector might also fulfil an important function in society when it comes to vision forming. Managers, especially in Brazil and Japan, could be important linking pins across levels of abstraction from theory to practice. In The Netherlands and Ghana, however, managers may need longer Extension to fulfil this role. Practical Workers were most similar across National Cultures, which is a beneficial characteristic if there is a need to develop consensus about the tasks at hand during the later stages of planning. The next step was to extrapolate the data from the empirical study to map similarities and differences in Time Perspectives on a global scale. The results of this theoretical step are described in the following chapter.

Chapter 5

Mapping international ambiguity in Problem Space

Andrew Segrave, Mariëlle van der Zouwen, Wim van Vierssen, and Bei Wen

At the start of any project that relies on intercultural cooperation it is essential to characterize how those involved frame the problems that the project is directed at. As described in the previous chapter, scientists tend to focus on less certain and longer term goals and problems than managers and practical workers in the water sector. And in the situation of a national culture that fosters virtues oriented towards future rewards these differences across professions are amplified. In this chapter the concept of Motivational Space is specifically applied to the framing of problems and is thus referred to as Problem Space. Data characterizing two dimensions of Problem Space in Ghana, Japan, Brazil, and the Netherlands were extrapolated to make a map of international ambiguity in Problem Space using conceptual and statistical relationships with Hofstede's National Culture Dimensions (Hofstede et al., 2010). The result may be used as an indicator of potential for intercultural cooperation on wicked water problems worldwide. And the aliens may be closer than anticipated: differences across professions in the water sector are greater than those across countries. Since ambiguity often impedes development insight into its location and type is valuable.

5.1 INTRODUCTION

5.1.1 Ambiguity in framing problems in the Water Sector

When faced with complex situations people unconsciously use interpretive lenses, or frames, to make sense of what is going on and extract what is important (Tversky and Kahneman, 1981; Elliott et al., 2003). Framing is the social construction of reality whereby individual and cultural characteristics influence how meaning is inferred, how values are attributed, and how people understand and respond to situations or issues. Framing determines what is considered at stake, who the stakeholders are, and how they behave (Nisbet, 2010; Brugnach and Ingram, 2012). The greater the differences in how problems are framed across groups the more difficult it is for those groups to communicate and cooperate successfully in working towards solutions (Elliott et al., 2003; Hulme, 2009). Ambiguity in framing is the simultaneous presence of multiple frames, which can make cooperation difficult (Bouwen and Taillieu, 2004; Dewulf et al., 2005; Pahl-Wostl et al., 2007b).

Ambiguity in framing is one factor that makes analysis and resolution of 'wicked' problems, such as those in the Water-Energy-Food-Climate nexus, so elusive (Rittel and Webber, 1973). In this contemporary example the idea of a nexus is used to emphasize a second impeding factor; that is, the interdependencies between various physical and social systems across multiple scales that make the problems complex (Hoff, 2011). On top of ambiguity and complexity, the work of planners is also afflicted with uncertainties: Planning depends on conceptualization of the future and the future inherently involves unknowns (Adam and Groves, 2007). As described in section 1.1.4, in response to wicked problems contemporary approaches to water planning strive toward ideal types that are intended to address these impeding factors (Table 5.1).

Table 5.1 Management responses related to properties of wicked problems.

Impeding factors	Ideal type response
Complex	Integrated
Uncertain	Adaptive
Ambiguous	Participatory

There has been much energy directed at achieving these ideal types, culminating in principles, like Integrated Water Resource Management, and policy instruments, such as the European Water Framework Directive. Practical implementation of these ideas, however, remains problematic. Participatory and democratic processes may be necessary (Rockenbach and Milinski, 2006; Buchan et al., 2008; Dreber and Nowak, 2008; Raihani, 2011), but inter-group cooperation fails when fundamental intercultural differences go unrecognized (Elliott et al., 2003). Developing insight into the enabling and constraining factors for intercultural cooperation on wicked problems is thus interesting and valuable.

5.1.2 Problem Space

In this chapter the concept of Motivational Space is specifically applied to the framing of problems and is thus referred to as Problem Space. The aim is to characterize the ambiguity in how different groups frame problems. Problem Space is defined by the Temporal Extent and Spatial Proximity of the objects that motivate individuals in a group to make decisions and act and the level of uncertainty that they attribute to these objects (Figure 5.1). At this point it is necessary to define these dimensions explicitly:

Motivational Object: a future goal or event that motivated an individual to make a decision or take action

Uncertainty: individual's subjective perception of the likelihood that Motivational Objects will occur

Temporal Extent: span of time populated by Motivational Objects; from the present into the future

Spatial Proximity: geographic scope of Motivational Objects relative to the individual

Rather than focus on the content of the specific objects that populate the Problem Space, such as goals, risks, and events, it is interesting to analyse these more fundamental, generic dimensions that define its scope. The main reason being: specific problems change while the Problem Space itself is more stable (Elliott et al., 2003). For a similar reason the focus was set on characterizing Problem Space at the group level and the researcher investigated the implications for intergroup cooperation. As with problems and Problem Space, it was assumed that the individuals in a group change frequently while groups themselves are more stable.

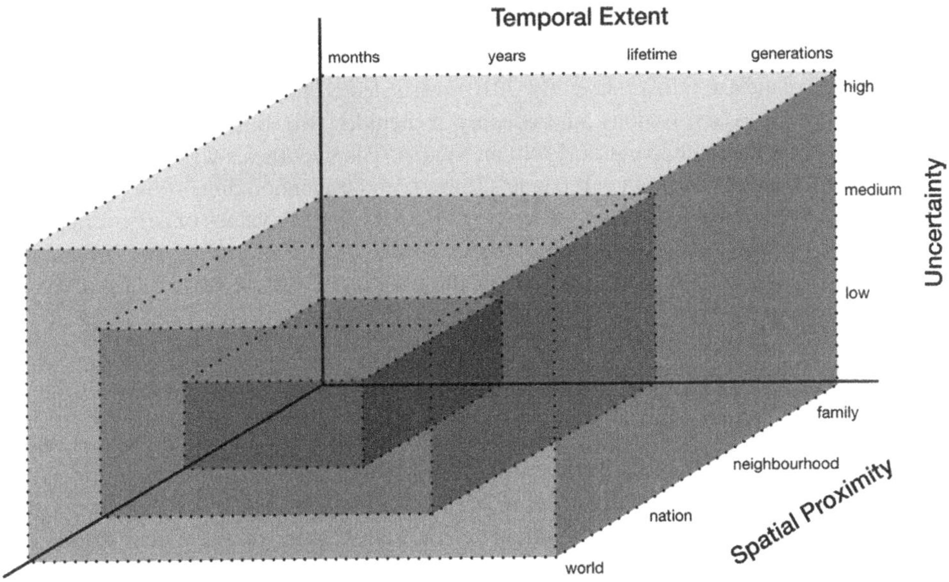

Figure 5.1 The dimensions of Problem Space.

Problem Space is a concept that has been used by various psychologists (Wallace, 1956, Piaget, 1969; Newell, 1980; Block, 1990), along with the related concepts of Life Space and Motivational Space (Graves, 1967; Lessing, 1968; Nuttin, 1984). The scope of the Problem Space as perceived by individuals in a certain group is seen as a product of framing and an expression of the culture of that group. For example, Boroditsky (2011) found that how people construct their mental representations of space and time depends on the languages they speak and the linguistic context. In this chapter the Problem Space of different groups is analysed to determine similarities and differences as an indication of potential for intercultural cooperation on wicked water problems worldwide.

5.1.3 Focus on the water sector

The water sector is an important and suitable case for investigating ambiguity in problem perception. Since water is a common pool resource that extends across national and administrative boundaries, water planning inherently necessitates transboundary cooperation between diverse groups. The Mississippi watershed, for example, extends across (parts of) 32 states in North America and two Canadian provinces. And the world's most international watershed, the Danube River Basin, includes the territories of 19 countries. Water issues are also a common thread in the main environmental challenges faced by societies today. It is for this reason that concepts such as 'the water footprint' and 'virtual water' have been developed to indicate the amount of water used by people in diverse processes and products (Allan, 1998; Hoekstra and Chapagain, 2007) and water is seen as the main interface linking systems in the Water-Energy-Food-Climate nexus (Hoff, 2011).

5.1.4 Aim and Hypotheses

For the theoretical study that is reported on in this chapter the aim was to map ambiguity in Problem Space on a global scale as an indicator of potential for intercultural cooperation on wicked water problems. The idea was to extrapolate data from the empirical study (Chapter 4) using correlations with measures of Cultural Proximity as defined by Geert Hofstede's National Culture Dimensions (NCDs). Hofstede's database (Hofstede et al., 2010) was selected for this purpose because it is global, with data for 76 countries and regions, and because it includes two dimensions that are conceptually similar to Temporal Extent and Certainty (Problem Space). The hypothesis was that Temporal Extent is conceptually and statistically related to Hofstede's Long-term Orientation dimension (LTO-CVS), which represents the importance that people from a certain culture attribute to values of perseverance and thrift, and that Certainty is conceptually and statistically related to Hofstede's Uncertainty Avoidance Index (UAI), which represents the level of anxiety that people from certain culture experience in the face of an uncertain future.

5.1.5 Focus on Temporal Extent and Certainty

The research depended on data from a previous empirical study that did not include Spatial Proximity. In this previous study cultural groups are differentiated according to national borders and professional roles to generate a representative sample of the variance in culture throughout the water sector worldwide. One-on-one interviews were then held with 309 Managers, Practical Workers, and Scientists working in Brazil, Ghana, Japan, and the Netherlands (Chapter 4). The objects that motivate the work-related decisions and actions of people in these groups were characterized in terms of Temporal Extent and Certainty. Temporal Extent was particularly central to this research for various reasons, as summarized below.

Environmental scientists and policy makers long ago theorized that most people are unlikely to perceive, let alone act on, problems on a long-term planetary scale (Meadows et al., 1972). And there is still much interest in risk perception (Giddens, 1999) and temporal discounting related to resource management (Mazur, 1987; Charlton et al., 2011; Bickel et al., 2012; Haith et al., 2012). Psychological research also suggests that behaviour is primarily motivated by short term objects (Okhuysen et al., 2003; Ainslie, 2005; Bickel et al., 2009), although considerable differences have been found on the individual scale (Zimbardo and Boyd, 2008). Cross-cultural studies have also revealed significant differences in time frames on a national scale (Gell, 1992; Levine, 1997; Raju, 2003). So people from different cultures conceptualize and deal with time quite differently while time and 'the future' are among the most essential concepts for planners.

The consequences of differences in the Temporal Extent used in framing problems are also evident in the decisions we make and the actions we take. In 2004, for example, when the Copenhagen Consensus Centre first ranked global policy measures by considerations of economic costs and benefits, solutions for diseases and malnutrition were given the highest priorities while those for climate scored the lowest (Lomberg, 2004). The costs of policy measures such as the Kyoto Protocol were seen to outweigh the benefits. The Stern Review reached a diametrically opposite conclusion just two years later (Stern, 2006). These two studies used similar economic cost-benefit analysis methods but reached contradictory conclusions because they made different value judgments about social costs and the rate of discounting the future (Hulme, 2009). Social discount rates are at the heart of cost benefit analyses. And for temporally extensive issues, such as Climate Change, the rate chosen basically determines the conclusions.

Problem Space is thus simplified into two dimensions: Temporal Extent and Uncertainty (Figure 5.2). In this hypothetical example, most people are motivated by relatively certain, short-term objects but on average people in Group 2 are motivated by more long-term, uncertain objects. This research project rests on the assumption that the framing of Problem Space is an expression of group culture that can be characterized on the individual level using psychological tests. The integrants of this assumption are specified in the sections below.

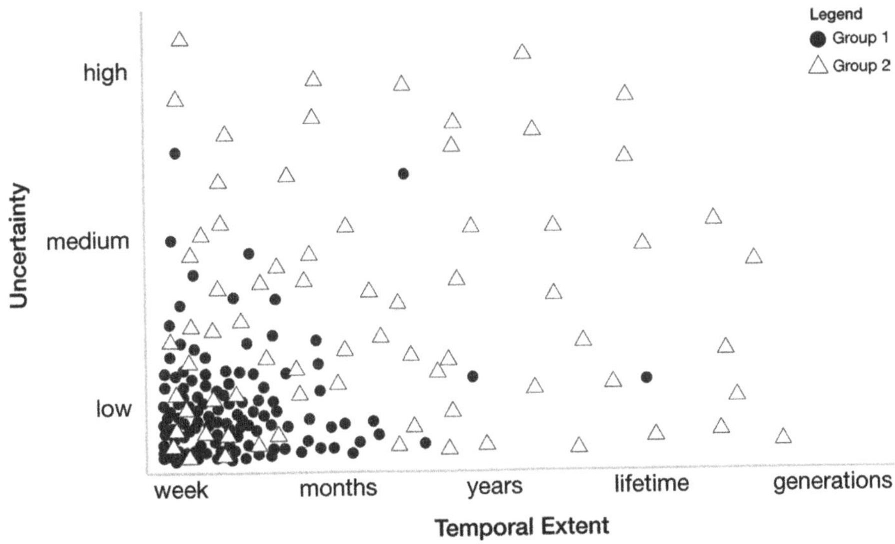

Figure 5.2 Simplified Problem Space populated with motivational objects.

5.1.6 Problem Space: an expression of group culture

Castells' (1996) vision of the Network Society introduced the idea that explosive growth of The Internet has made networks the new social morphology of our societies. In a networked society individual actors are still the basic unit, but now they are linked via networks to others all over the globe. These networks have transformed the nature of cooperation and raised fundamental questions about what constitutes a group or a team (Wageman et al., 2012b). Group culture is taken to mean the characteristics of a certain group of individuals that makes the group identifiable. A social group may be small and virtual, such as an online community, or large and geographically bounded, such as a nation. In both cases the group has certain distinguishing characteristics, such as language, clothing customs and expectations concerning behaviour, which also change albeit mostly more slowly (Geertz, 1973).

Individuals develop, sustain, and transmit the culture of the group to future generations. And this reinforcement of culture over time occurs through interaction with the heterogeneous network within which the individuals exist, be that network virtual or otherwise (Barney, 2004). Most individuals are simultaneously members of multiple groups, which may be embedded in each other, overlapping, or completely orthogonal (Roccas and Brewer, 2002). While regional cultural differences and dynamics are recognized, national borders were accepted as rough indications of divisions between cultures on a global scale as have previous studies (e.g. Enserink et al., 2007). Besides this crude dissection, three sub-groups based on professional roles were also analysed: Practical Workers, Managers, and Scientists. These groups were chosen to represent the variation across different professions in the water sector as regards the level of abstraction from theory to practice.

5.1.7 The individual as atom, with quarks

The individual is held to be the most clearly autonomous entity, and thus agent, in any social system. Individuals are seen as windows into group culture. Even so, each individual has multiple identities associated with the various roles they perform (Goffman, 1959; Schlenker, 1985; Ting-Toomey, 1993).

An individual may alternate between being father, brother, sportsperson, and scientist - all in one day. And the manifestation of each identity is generated in a heterogeneous network as an outcome of the interaction between the individual and the context. Interaction between individuals happens by a process of 'identity negotiation' whereby they establish mutual expectations of one another to form groups (Swann, 1987; Cinyabuguma et al., 2005). The identities of an individual are "contested, contextual, multiple, fragmentary and transient" (Barney, 2004, p.17), and so one individual may provide insight into various group cultures. This is why it is important to speak with an individual in the context of the specific group in question (Chapter 3).

5.1.8 Cross-cultural comparison: from what perspective?

Culture has been the subject of much research from a diverse range of perspectives and based on various, often conflicting, ontological, epistemological, and methodological assumptions. Psychologists, sociologists, anthropologists and historians may attribute quite different meanings to the word 'culture', but there are also considerable inconsistencies within each discipline. One widely used approach for analysing cultures is to compare traits regarding attitudes and values, which are often simplified as dualisms such as individualism and collectivism, across individuals associated with identifiable groups (Triandis, 1996; Oyserman et al., 2002; Coon, and Kemmelmeier, 2002.). Minkov (2012) recently examined the theoretical and practical advantages and potential pitfalls of this type of cross-cultural analysis.

There has been criticism of past cross-cultural comparisons from a methodological perspective (e.g. Heine et al., 2002), due to problems concerning language issues (e.g. Bontempo, 1993; Kashima and Kashima, 1998), construct equivalence (e.g. Norenzayan, et al., 2010), and cross-cultural differences regarding how questionnaires themselves are treated (e.g. Peng et al., 1997). In response, Hambleton et al. (2005) developed guidelines for translating and adapting psychological tests for cross cultural assessment. But this does not deal with the more fundamental criticism of the ontological assumptions upon which this form of psychologism rests (Shore, 1996). Premises about human agency and social structure and the relationships between them are, for example, much disputed (Lee and Brown, 1994; Burke, 2005).

It is necessary to locate the author's perspective in this wide range of approaches (Guba and Lincoln, 2005). Critical Realism offers ontological and epistemological grounds that are most consistent with those taken in this book (Chapter 2: Interpretive Framework). Archer developed the Morphogenetic Approach to analysing social systems, which fits into this tradition. Based on systems theory, the Morphogenetic Approach holds that the properties which characterize socio-cultural systems emerge from the complex interaction between social structure and human agency. By retaining the distinction between agency and structure, this approach provides insight into their interaction (Archer, 1982). The author understands that real systems may be continua without actual categories, such as actors, networks, and structure, but as did Archer the analytical utility of such constructs is also recognized. Archer's model of social change, being the product of the relative strength of both voluntary and deterministic forces, seemed to provide the most coherent grounds for a study of Problem Space.

5.2 METHOD

To test the hypotheses (section 5.1.4) a theoretical inquiry was supported by statistical analysis following four main steps as explicated below.

Step 1: Assess the theoretical equivalence of Hofstede's National Culture Dimensions (NCDs) and the measures of Problem Space

The definitions and underlying rationale behind the 6 NCDs was taken from the most recent print of Hofstede's book (Hofstede et al., 2010). Deeper meanings and assumptions were also considered based on Minkov's book about cross-cultural analysis (Minkov, 2012). These concepts were compared with those used in designing a multi-measure interview method for an empirical study in Brazil, Ghana, Japan, and the Netherlands (Chapter 4).

Step 2: Statistical analysis of the correlations

To compare Hofstede's national scale data with the individual level data, which did not fit a normal distribution when grouped per country, national medians were calculated for Temporal Extent and Certainty. Spearman's rank correlation coefficients were then derived using these values and data for the NCDs that were theoretically most equivalent (Step 1). The main limitation to this theoretical analysis was the small sample size for Temporal Extent and Certainty on the national scale (n = 4). NCD data was not available on the sub-scale of professional roles or on the individual scale. Empirical data from both studies was available for Japan, Brazil, and the Netherlands. Not all of the countries in Hofstede's study were characterized using all 6 NCDs. For Ghana it was necessary to predict a value for Hofstede's original measure of long-term orientation (LTO-CVS) using the larger dataset that was based on the World Values Survey (LTO-WVS) and the strongest correlations with the other NCDs (Section 5.3.1) National values for all other NCDs were obtained directly from the Hofstede's study.

Step 3: Map Cultural Proximity worldwide and evaluate the effect size of the measures of Problem Space

This step only included the NCD data. Cultural Proximity was defined as the product of the similarities and differences across all six of these NCDs:

1. Power distance (PDI): the degree to which inequalities in the distribution of power are expected and accepted
2. Individualism (IDV): the degree to which individual autonomy and responsibility is preferred over interdependence and loyalty in a close-knit social framework
3. Masculinity (MAS): the degree to which achievement, heroism, assertiveness and material rewards are preferred over cooperation, modesty, caring for the weak, and quality of life.
4. Uncertainty avoidance (UAI): the degree to which the members of a culture feel threatened by ambiguous or unknown situations
5. Long-term orientation (LTO): the degree to which virtues oriented towards future rewards are fostered. Note: In all calculations the original LTO-CVS data was used instead of the later derived LTOWVS data, which is conceptually further removed from Temporal Extent (See Results Step 1).
6. Indulgence versus restraint (IVR): the degree to which basic and natural human drives are regulated

Data on these 6 dimensions was available for 66 countries, which formed a 66-by-6 matrix. The multidimensional scaling (MDS) approach was used to visualize Cultural Proximity between all 66 countries in a two-dimensional Euclidean space. On the resulting global map the position of each country relative to others indicates the sum of the differences across all 6 NCDs. To make this map a symmetric 66-by-66 dissimilarity matrix was first created, where the Euclidean distance between each pair of countries was derived based on the 6 NCDs. MDS operates on this distance matrix to reduce the original data into two dimensions, which were then visualized in a two-dimensional scatter plot.

The K-means clustering algorithm was applied to the rescaled data for partitioning the countries into an arbitrarily chosen number of clusters (K = 5). The distance matrix, MDS analysis, and K-means clustering were conducted using R-functions 'dist', 'cmdscale', and 'kmeans', respectively. Lastly, clusters were inspected and interpreted. Factor analysis was used to evaluate the effect size of the NCDs (LTO and UAI) that were theoretically and statistically related to Problem Space (Temporal Extent and Certainty). In this way international similarities and differences based on Problem Space could be interpreted in the broader context of Cultural Proximity.

Step 4: Map international ambiguity in Problem Space

The 22 countries were first mapped according to LTOCVS and UAI scores and clustered into four types of 'NCD Problem Space' using multinomial logistic regression based on all possible values: high LTOCVS and high UAI; high LTO-CVS and low UAI; low LTO-CVS and high UAI, and low LTO-CVS and low UAI. Since Spearman's coefficients were derived using a limited number of countries (n = 4), it would be inappropriate to extrapolate individual data points for the 22 countries. For this reason the four types of Problem Space (shaded areas) were extrapolated instead.

Two prediction models were prepared using linear regression ('glm' function in R) to extrapolate the empirical data for all values within the ranges of the four types of 'NCD Problem Space' (model 1: Temporal Extent regresses on LTO-CVS; model 2: Certainty regresses on UAI). This resulted in a map of international ambiguity in Problem Space in terms of Temporal Extent and Certainty with regions indicating where the countries grouped by the associated LTO-CVS and UAI types are likely to be located. The countries for which empirical data was available (Ghana, Brazil, Japan, and the Netherlands) were also indicated. Finally, data points were plotted for three subgroups based on Professional Roles (Practical Workers; Managers, and Scientists) to reveal the significance of the differences in Problem Space across professions within the context of international ambiguity.

5.3 RESULTS

The results are organised following the four steps of the method.

5.3.1 The theoretical relationship between the measures

Based on Hofstede's explanations and premises it is first reasoned that Temporal Extent is conceptually related to Long Term Orientation. Secondly it is theorized that Certainty is positively related to the Uncertainty Avoidance Index.

Temporal Extent is related to Long Term Orientation (LTOCVS)

Hofstede defined LTO-CVS as "the fostering of virtues oriented towards future rewards – in particular, perseverance and thrift." While Short Term Orientation was defined as "the fostering of values related to the past and present – in particular, respect for tradition, preservation of 'face', and fulfilling social obligations." Analysis of Hofstede's explanation of LTO-CVS revealed that high scores also represent certain Confucian values such as adaptiveness, owing to its origins in the Chinese Value Survey, and value of (situation dependent) virtue over (absolute or universal) truth or good. This is likely to explain why five Asian countries score highest on this dimension. In practice (Hofstede et al., 2010), the LTO-CVS dimension was characterized using two divergent sets of values (Table 5.2).

Table 5.2 Values associated with LTO-CVS.

Low LTO combined:	High LTO combined:
Reciprocation of greetings, favours, & gifts	Persistence (perseverance)
Respect for tradition	Thrift
Protecting one's "face"	Ordering relationships by status and observing this order
Personal steadiness and stability	Having a sense of shame

One limitation of the LTO-CVS dataset is that it only includes 23 countries. Minkov later used World Values Survey data to increase the number of countries characterized. This meant redefining the dimension (LTO-WVS) using scores based on three items (Hofstede et al., 2010, p. 253):

1. Thrift as a desirable trait for children
2. National Pride
3. Importance of service to others

Even though LTO-WVS correlates strongly with LTOCVS ($r = 0.72$), there are some significant discrepancies between the two datasets. Since the original LTO-CVS data is based on empirical results from a purpose designed questionnaire preference was given to these values over the statistically derived LTO-WVS data.

In assessing the theoretical relationship between Longterm Orientation and Temporal Extent it is interesting to note that Hofstede found deferment of gratification to vary with LTO-CVS. People from countries that score highly on the LTO-CVS scale are more likely to perceive and choose long-term benefits over short-term benefits (Hofstede et al., 2010, p. 241). Hofstede also found that High LTO-CVS correlated with people the perceived importance of profits 10 years from now and the unimportance of this year's profit (Hofstede et al., 2002). This shows how values such as thrift and perseverance are likely to influence the nature of the goals and problems that people perceive. The hypothesis is that these values also manifest themselves in the Temporal Extent of Problem Space because, as further described in the introduction, Temporal Extent represents the time quality of the objects that motivate people (Nuttin, 1984).

Certainty is related to Uncertainty Avoidance Index (UAI)

UAI is defined as "the extent to which the members of a culture feel threatened by ambiguous or unknown situations" (Hofstede et al., 2010, p. 190). This dimension was characterised using scores on three items:

1. Level of job stress
2. Rule orientation
3. Intent to stay at the current workplace

Hofstede took these three items to be "expressions of the level of anxiety that exists in a particular society in the face of an uncertain future" (Hofstede et al., 2010, p. 191). People who want to avoid uncertainty seem to value rules and rituals. The link was also made with Maslow's hierarchy of human needs, which places 'safety and security' under 'belongingness', 'esteem', and 'self-actualisation'. Hofstede argues that 'safety and security' are likely to prevail over other needs where uncertainty avoidance is strong, and that differences in UAI imply differences in motivation patterns - especially if combined with other dimensions such as Masculinity (Hofstede et al., 2010, p. 213). This conclusion is particularly interesting for the study of Problem Space as framed around Motivational Objects. People who see the uncertainty inherent in life as a continuous threat that must be fought are, in theory, less likely to set goals for themselves or allow for events that they perceive to be highly uncertain. In theory, groups that score highly on UAI will be motivated by objects that are attributed with greater Certainty.

5.3.2 Statistical analysis of the correlations

In psychological research, Cohen's (1988) conventions are used to interpret effect size. A correlation coefficient of .10 represents a weak association; .30 indicates a moderate correlation; and .50 or larger is a strong correlation. Using the national level data for Japan, Brazil, Netherlands, and Ghana a significant

($p < 0.05$) and strong ($r > .50$) correlation was found between Temporal Extent and LTO-CVS ($r = 0.99$, Figure 5.3).

This was also true for Certainty and UAI ($r = 0.99$, Figure 5.4). Separate analysis of the sub-groups scientists, managers, and practical workers also revealed significant ($p < 0.05$, except for practical workers) and strong ($r > .50$) correlations (Figures 5.3 and 5.4). Scientists are generally motivated by more temporally distant objects, while managers attribute more Certainty to their MOs.

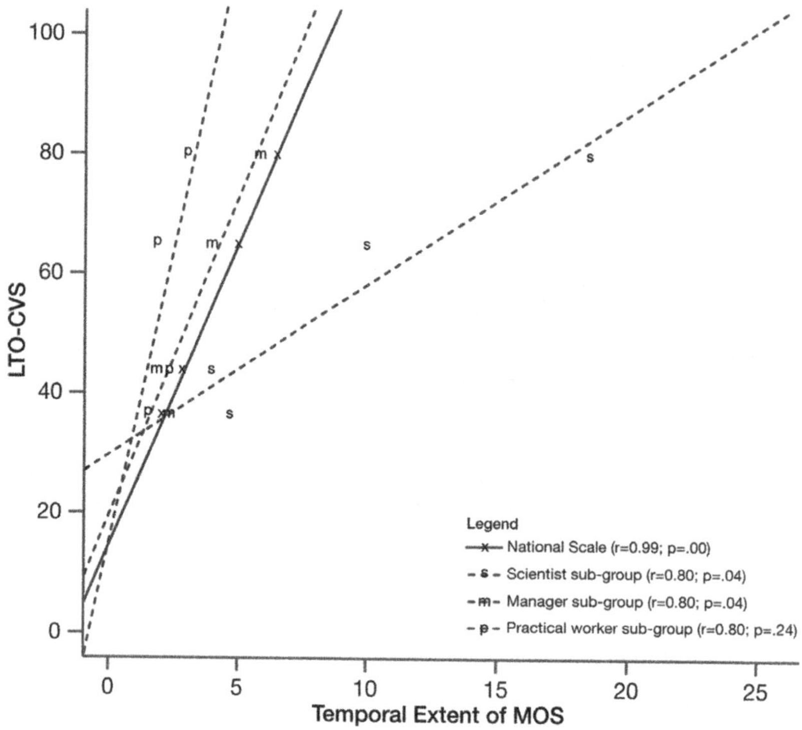

Figure 5.3 Correlation between Temporal Extent of MOs and LTO-CVS.

5.3.3 Cultural Proximity worldwide & the effect size of the measures of Problem Space

The result of mapping Cultural Proximity between each pair of 66 countries is a 2-dimensional MDS map (Figure 5.5). Countries are clustered into five groups based on the C-means clustering technique. After inspection, these clusters were interpreted and named based on shared characteristics across the 6 NCDs for the modes in each cluster. The main distinguishing factors were (1) individualism versus (2) collectivism and (3) long-term versus (4) short-term orientation. The fifth cluster includes cultural 'intermediates' that are unexceptional on any of the NCDs. These clusters are not reducible to a single NCD, however, so the cluster names are slightly simplistic.

Factor analysis (FA) was conducted to better understand the distribution of countries on Figure 5.5. PDI and IDV are highly correlated with Principle Component (PC) 1 ($r_{PDI} = .88$; $r_{IDV} = -.87$), while LTO and IVR are correlated with the PC2 (r_{LTO} = .89; r = -.81). Significant and strong correlations were a found between the PC1 and the MDS Dimension 1 (x-axis) values and between PC2 and the MDS Dimension 2 (y-axis) values. This demonstrates high consistency between the two different dimension reduction methodologies (FA and MDS). PDI and IDV are highly correlated with Dimension 1 (r_{PDI} = .84; r_{IDV} = -.76), while LTO and IVR indices were correlated with Dimension 2 (r_{LTO} = .84; r_{IVR} = -.64). A negative correlation exists between PDI and IDV, which is also true for LTO and IVR. Hofstede discusses the theoretical relationships between different pairs of NCDs in his book and Figure 5.5 is consistent with these findings (Hofstede et al., 2010).

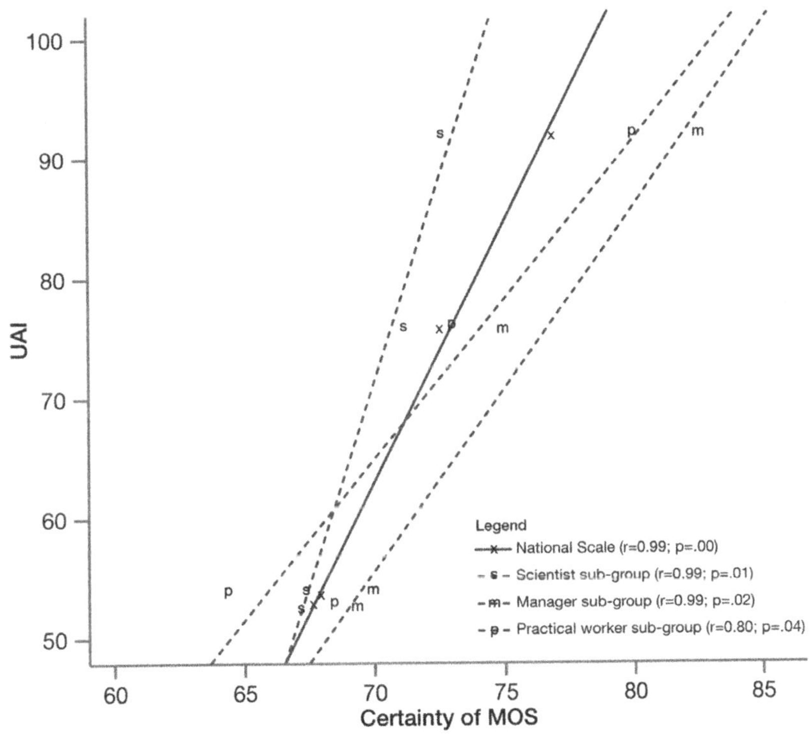

Figure 5.4 Correlation between certainty of MOs and UAI.

Looking closer at the determinants of Cultural Proximity the two principal components, PC1 (PDI & IDV) and PC2 (LTO & IVR), explain most of the total variance across the 6 NCDs (30% and 25%).The other four PCs together explain over 40%. MDS 1 explains 57% and MDS 2 43% of the total variance across the 6 NCDs. In other words, the variance across countries in terms of PDI and IDV combined is greater than those in terms of any other combination of dimensions. Even so, LTO, and by correlation Temporal Extent, is the strongest single variable determining the Cultural Proximity of countries (r = .84). LTO also spreads the countries in a different direction (x-axis) to the other dimensions (y-axis), indicating that it is a relatively unique variable. On the other hand UAI, and by correlation Certainty, has relatively less influence on Cultural Proximity.

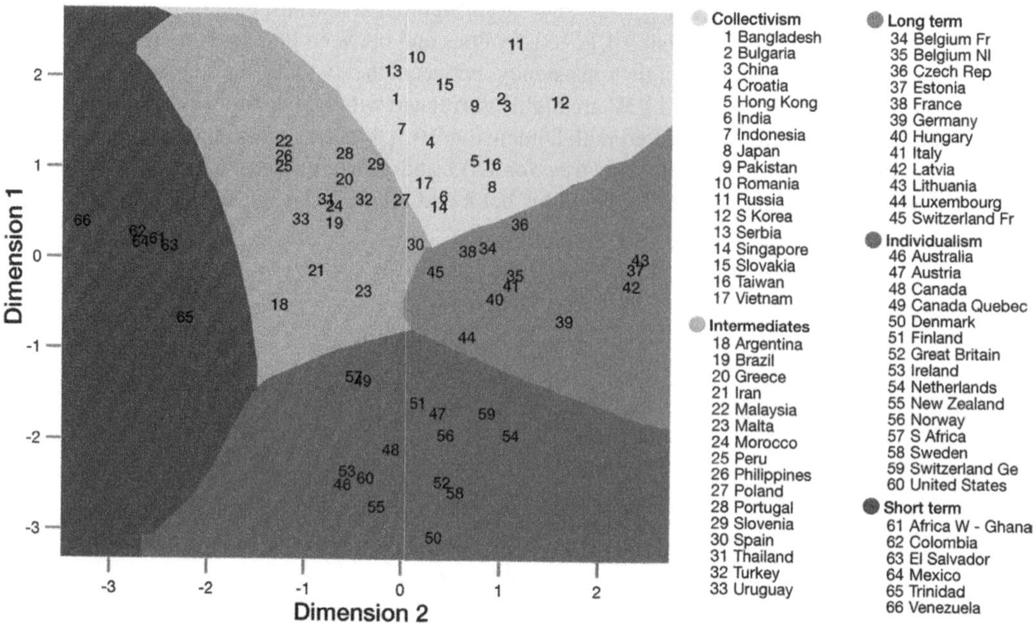

Figure 5.5 Map of Cultural Proximity for 66 countries.

5.3.4 International ambiguity in Problem Space

In Figure 5.6 the map of Cultural Proximity has been simplified into 'NCD Problem Space'. Plotted on this map are the 22 countries for which LTO-CVS and UAI data was available. The shaded areas represent four types of Problem Space: high LTO-CVS and high UAI; high LTO-CVS and low UAI; low LTO-CVS and high UAI, and low LTO-CVS and low UAI.

In Figure 5.7 Problem Space is mapped in terms of Temporal Extent and Certainty. The shaded areas indicate the four types of Problem Space that were extrapolated based on correlations between Temporal Extent and LTO-CVS (x-axis), and between Certainty and UAI (y-axis). Plotted in this Problem Space are the 4 countries for which empirical Temporal Extent and Certainty data was available. Subgroups based on Professional Roles are plotted in addition to these national scale data points.

5.4 DISCUSSION

Theoretically defensible reasons are given for comparing LTO and UAI with Temporal Extent and Certainty (Step 1). Even so, the number of national scale data points was limited ($n = 4$). Taking this limitation into consideration, the correlations support the hypothesis more strongly than expected with Pearson's coefficients (r) approaching +1 (Figures 5.3 and 5.4). The weakest correlation is between Temporal Extent and LTO-CVS for the Practical Worker subgroup ($r = 0.76$). There may be a functional limit to the temporal extent of what motivates Practical Workers in any cultural setting, since their decisions and actions have physical outcomes in the present. All the same, even this correlation forms a significant and strong enough basis for extrapolation.

In addition to the remarkable strength of the correlations, the gradients are particularly interesting: higher LTO-CVS and UAI scores are associated with greater differences between Professional Roles. This

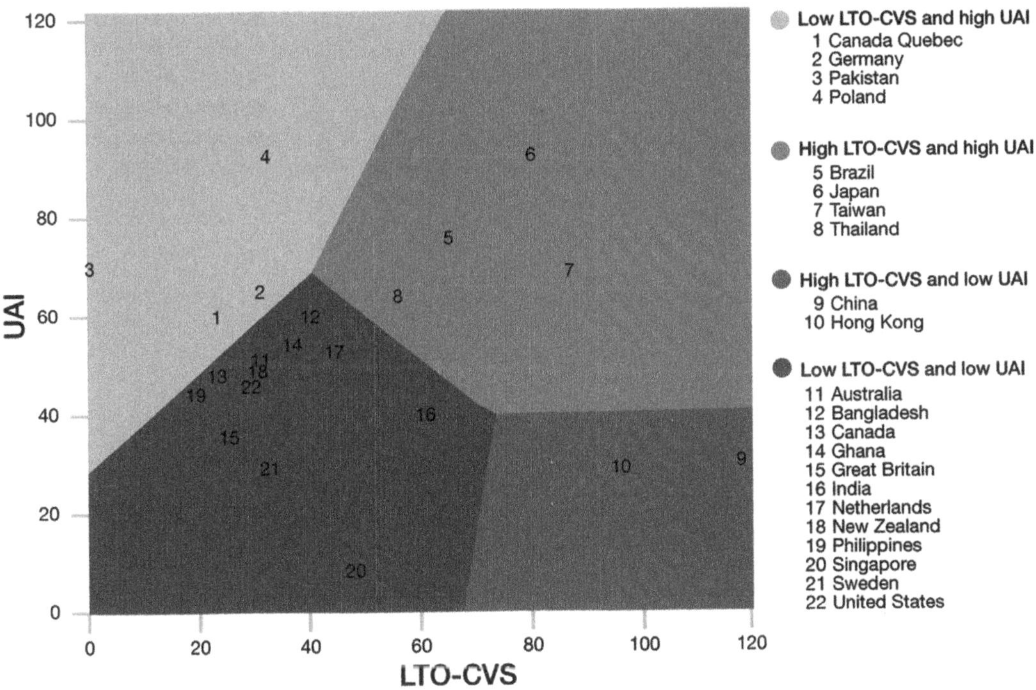

Figure 5.6 NCD Problem Space UAI.

pattern is especially strong for Scientists. In countries characterized by high LTO-CVS scores, Scientists tend to be motivated by markedly longer term problems and goals (greater Temporal Extent). It seems that a cultural tendency towards, and perhaps acceptance of, long-term orientation may allow Scientists to explore more distant time horizons.

Hofstede found that for cultures which are more comfortable with uncertainty "time is a framework in which to orient oneself but not something one is constantly watching" (Hofstede et al., 2010, p. 210). This freedom from the clock may explain, to some extent, the long term perspective of Scientists in Brazil and Japan (Ylijoki and Mäntylä, 2003). Perhaps functional limits to Temporal Extent for water professionals lie further into the future than those emerging through identity negotiation within a certain national culture; especially for scientists whose work is inherently exploratory. Scientists in the Dutch water sector did tend to measure their worth in terms of practical applicability and reliability. In contrast, some Brazilian scientists seemed to be proud to be seen as an 'oracle' or source of wisdom.

As with LTO-CVS, higher UAI scores were associated with greater differences between professions as regards Certainty. These patterns are reflected in the final map of international ambiguity in Problem Space (Figure 5.7). Ambiguity in Problem Space across professional groups may be at least as problematic for cooperation on wicked water problems as that across national cultures. This begs the question: who are the aliens actually? These results suggest that Problem Space needs to be dealt with explicitly in any projects that require cooperation across professional roles - even on a subnational scale. When it comes to international scale it is also important to view ambiguity in Problem Space within the context of Cultural Proximity.

The distribution of countries on the map of Cultural Proximity (Figure 5.5) is validated by high consistency across the different methods used (PCA and MDS). The international similarities and differences in terms

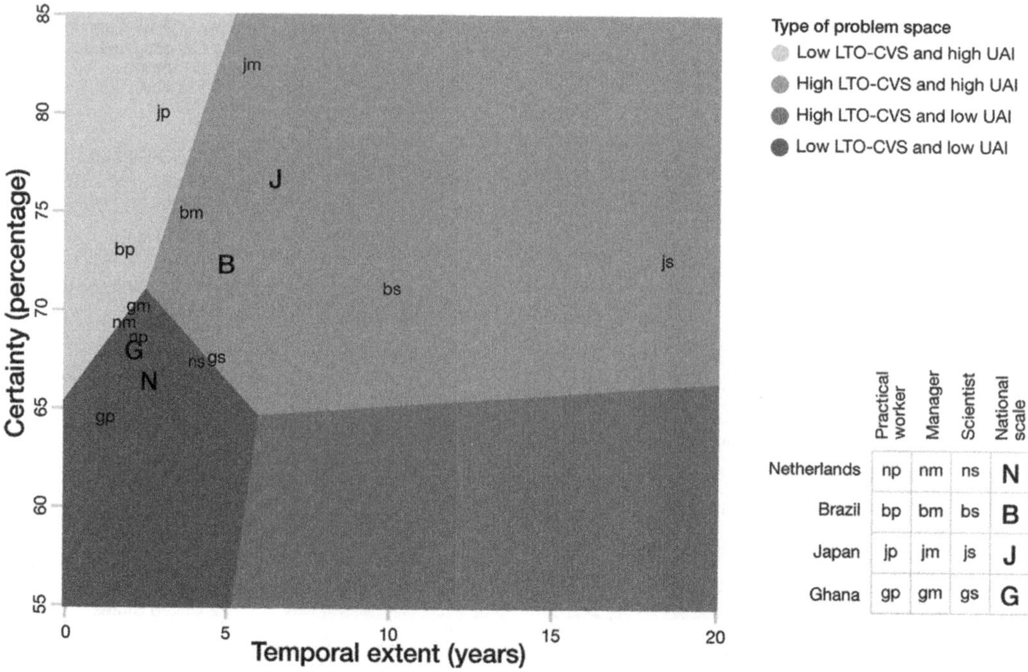

Figure 5.7 Map of international ambiguity in Problem Space for the water sector.

of LTO-CVS were found to be large and distinctive within the broader context of Cultural Proximity. The diversity in terms of PDI, IDV & IVR is, however, not markedly less significant and diversity in terms of UAI is less pronounced. This result confirms that Problem Space is a meaningful and useful construct, but also indicates that explanation of the international ambiguity in Problem Space should take the influence of PDI, IDV & IVR into consideration.

Although the map of NCD Problem Space (Figure 5.6) is an intermediary result, it also shows international ambiguity in Problem space with empirically derived data points. It is interesting to note that Brazil, India, and China, countries whose answers to wicked problems in the Water-Energy-Food nexus will be of global importance, are likely to frame problems quite differently from each other but, more importantly, with a longer term vision than recent world powers such as the United States of America, the United Kingdom, and Germany. The added value of the final result (Figure 5.7) is that the Problem Space is defined by scalar dimensions, with units in years and percent, and that sub-national groups of Professional Roles are also plotted specifically for the water sector.

5.5 CONCLUSIONS

The hypotheses were confirmed with theoretical and statistical substantiation, which made it possible to map ambiguity in the framing of Problem Space as an indicator of the potential for intercultural cooperation on wicked water problems worldwide. International differences in Problem Space were also evaluated relative to a more general map of Cultural Proximity. This indicated that Long-term Orientation, and by correlation Temporal Extent, had relatively strong influence on the Cultural Proximity of countries whereas Uncertainty Avoidance, and by correlation Certainty, had less influence. These results show that ambiguity in the framing of Problem Space is likely to be an important aspect of cross-cultural differences, especially as regards Temporal Extent.

Problem Space was mapped, first using the NCDs and then via correlations in terms of Temporal Extent and Certainty. These maps show similarities and differences between 22 countries as to how Problem Space is likely to be framed. Ambiguity in the framing of Problem Space has thus been roughly characterized on a global scale. The results revealed a perhaps lesser known face of ambiguity: differences in Problem Space are greater across professional roles in the water sector than across national cultures. The international differences also seem to amplify or diminish differences across professional roles - especially for scientists.

Ambiguity surrounding problem perception is likely to have far reaching consequences for how the future of the water sector is being envisioned and formed. Which subjects dominate research agendas? What projects are prioritized by practitioners? How do we achieve international collaboration in response to wicked problems such as Climate Change and loss of biodiversity? Managers and practical workers in the water sector are motivated by more certain, short-term objects and thus frame the Problem Space quite differently to scientists. This ambiguity is likely to explain, at least to some extent, the difficulties experienced in translating wicked water problems signalled by scientists into practical response strategies. Attention should thus be given to explicitly framing Problem Space for all such projects.

A first step in the implementation of innovation theories and frameworks that rely on collaboration across roles should be to characterize the ambiguity in the framing of the problems that the innovation initiative is directed at solving. Furthermore, a collaborative entrepreneurial innovation model may work better in countries that are orientated towards the shorter term, but may also further inhibit scientists in their freedom to explore (Figure 5.6). This leads us to suggest a new role for scientists in helping societies to become more sustainable: one of futures research and agenda setting… the scientist as oracle. It would be fascinating to focus future research on more detailed comparisons across other sub-national groups and perhaps include the spatial dimension, which could generate more specific insights to help water planners cooperate in practice.

Chapter 6

Conclusions, implications and applications

Andrew Segrave, Mariëlle van der Zouwen and Wim van Vierssen

The foreseeable future - what a concept! Commonly used in water planning, policy making, and even in environmental law (Gell, 1996), the concept is unclear and an important source of miscommunication between key actors in the water sector. The future is inherently unknown, so what do the actors believe can be foreseen? Chapters 4 and 5 describe how the answer to this question varies significantly across cultures on a national level, but even more so across professions in the water sector. Some people are certain their ten-year plans will be achieved, while others doubt and wonder about tomorrow. These cross-cultural differences may be remarkable, but they are also critical for how people make decisions and act.

The time horizons used for framing problems and visions are critical for setting research agendas, water governance, and planning transitions to the cities and river basins of the future because they basically determine the questions asked and which solutions are sought. The future objects identified by 309 water professionals interviewed in Brazil, Ghana, Japan, and the Netherlands as being most important, were characterized in Chapter 4, revealing remarkable differences in the temporal extent of the goals and events that motivate people. Empirical data from these 4 countries was extrapolated in a subsequent theoretical investigation to map the key characteristics of Motivational Space for 22 countries (Chapter 5: Mapping international ambiguity in Problem Space). This theoretical study resulted in an indicator of potential for worldwide intercultural cooperation on water problems.

The temporal extent of the objects that motivate people's actions is, in general, considerably shorter than the time span associated with the (unintended) consequences of their actions. Scientists are motivated by objects that lie furthest into the future, followed by managers and then practical workers (Group medians: 8.4, 3.5, 1.7 years). Adam and Groves (2007) make a strong ideological argument against this mismatch, listing fossil fuels, plastics, nuclear energy, and geno-technology as examples of 'institutional irresponsibility'. But what can individuals and organisations do about this now? It is tempting to join the criticism of current ways of thinking and define a fourth nirvana concept for water managers: adding 'long-term' to the 'adaptive', 'integrated', and 'participatory' ideals (Molle, 2008; Pahl-Wostl et al., 2007a).

In concluding his analysis of societal collapse Wright (2004) warns that what is needed, above all, is a transition from short-term to long-term thinking. If the author were only to confirm this conclusion, however true it may be, it would add little to the existing knowledge and this line of thought is also unlikely to be practically helpful. For this reason a more pragmatic approach is adopted in this last Chapter, which explores some possible consequences of the conclusions. The objective of this Chapter is to discuss the implications and applications for (1) water planning and (2) setting water research agendas. The conclusions include practical recommendations but the results also have wider implications. Some lines of thought for future studies are thus also explored using ideas from complexity science and Adam and Groves' (2007) call to rebuild the dependencies between action, knowledge and ethics when dealing with the future.

The structure of this final chapter of the book is facilitated by the original six research questions presented in Chapter 1. The chapter begins with a section that is dedicated to answering the first three research questions, which are about the variables and the method. The next section focusses on questions four and five, which are about the empirical results concerning ambiguity in Time Perspectives. Two sections are dedicated to answering question six, which is about the implications and applications of the empirical findings, because these conclusions depend on the scale considered. First all of the international ambiguity in Motivational Space is taken into consideration, including the similarities and differences across all National Cultures and Professional Roles. Next the focus is set on the ambiguity in Motivational Space across Professional Roles on a sub-national scale and for different types of countries. The investigation of implications and applications rested on and raised various theoretical issues, which are reviewed in the theoretical discussion that follows these conclusions. The Chapter is concluded with a list of recommendations.

6.1 CONCLUSIONS

It is interesting to ponder the fact that many philosophers have rejected the existence time, or some aspect of it. The influence of various concepts of time on how people think and act is, however, very real. As described at the outset of this research project, the aim was to typify how people working in the water sector worldwide conceptualize and deal with time and to map the ambiguity in Time Perspectives by characterizing the similarities and differences across groups. A set of six sequential research questions was defined in support of achieving this goal. The answers to these questions, which were developed, described and substantiated in the previous Chapters, are presented below to conclude the structured analysis and set up a springboard for exploring some broader implications and applications.

6.1.1 Variables and method

The first three research questions were directed at the definition of variables and the choice of a method. A brief discussion about the value of the methodological advancements made during the project is included because they represent fundamental steps that are sure to be useful to other researchers.

1. Which aspects of an individual's Time Perspective have influence on the decisions they make and the actions they take?
2. How do the relevant aspects of an individual's Time Perspective relate to each other?
3. How can the relevant aspects of an individual's time perspective be characterized or measured in an integrated fashion?

One innovation that was conceived and applied in this research project was the adaptation and integration of previously distinct measures into a new multi-measure method that is based on explicit ontological and epistemological premises and a model of Time Perspective. This new method (Chapter 3) represents the comprehensive answer to Question 3 – what follows is a summary. Designing and defending a new method and model, rather than applying a ready-made and widely accepted one, was daunting but deemed necessary. Existing measures could be used to typify certain dimensions separately but all of these dimensions are interrelated and are likely to influence decisions and actions both directly and indirectly.

As described in section 2.3: Conceptual Framework, the focus of the empirical investigation was finally set on characterizing the variables on the scale of individuals. Besides the methodological criteria related to the research questions, this focus derives from several ontological and epistemological considerations as described in section 2.1: Ontological position and model of Time Perspective. Put briefly, since a Time Perspective is a cognitive representation of actual time it is only be made real by the influence it has on how someone thinks or acts. The implications are significant. Motivational Objects, as first person accounts of the 'what' given in reasons 'why', were analysed to reveal fingerprints of that person's Time Perspective. Besides the Temporal Extension and Certainty attributed to these objects, this analysis also included characterizing the more fundamental dimensions of Structure and Orientation. Other dimensions of Time Perspective, such as Content, Density, etc., were assumed to represent more detailed characteristics of Motivational Objects and were not included as distinct dimensions.

Similarities and differences across groups were identified using the Foreseeable Future Multi-measure Method in structured interviews (Chapter 3). This method was designed based on the idea that all aspects of an individual's Time Perspective have influence on the decisions they make and the actions they take. This is also a short answer to Question 1. Some aspects vary more than others across groups but this says nothing about the relative magnitudes of influence. The study did not include empirical observations of decisions or actions but rather memories and plans for these as reported by the interviewees. It would be interesting and valuable to augment this research project with a longitudinal, empirical study of the actual decisions and actions made in the everyday working lives of those interviewed.

The answer to Question 2 requires slightly more explanation: Theoretical relationships between the various dimensions of Time Perspective are included in a model described in the Interpretive Framework (Chapter 2) as a basis for defining the variables and designing a method. The dependent variable, Foreseeable Future, was defined as the Temporal Extension of an individual's Motivational Objects and the Certainty attributed to these objects. Temporal Extension is relative to how the Structure of time is conceptualized. It is a measure of the magnitude of the interval between two points; that is, 'present' and 'future' moments located in an envisaged model of time that is itself variable. Orientation towards or against the future relative to the past and the present is also likely to colour the way in which the other dimensions of Time Perspective are conceptualized. The relationships between variables are discussed in more depth in Chapter 4.

From a statistical perspective, no correlation was discovered between Temporal Extension and the number of Motivational Objects reported, or between Temporal Extension and the Certainty attributed to these goals and events. On the other hand, an interesting relationship was identified between Temporal Extension and Structure: Groups characterized as having a longer Temporal Extension into the future also tended to perceive the present as having the shortest duration relative to the past and the future. Groups characterized as having a shorter Temporal Extension into the future perceived the past as comprising less time than the present. Although Temporal Extension into the past was not quantified in this study, these results seem to indicate that extension into the future is positively related to extension into the past. It is also interesting to note that no statistical relationship was found between Orientation and Temporal Extension. As reported in Chapter 5, a statistically significant and strong correlation was found between the data for Temporal Extension and Certainty and that for Hostede's (2010) national scale cultural dimensions of Long-Term Orientation (LTO-CVS) and Uncertainty Avoidance (UAI). Differences across groups on a national scale amplify those across professions. More similarities and differences across groups are described in the next section.

6.1.2 Ambiguity in Time Perspectives

The fourth and fifth research questions were aimed at gaining insight into the similarities and differences across groups in terms of Time Perspective:

4. To what degree are the Time Perspectives of people working in different professions throughout the water sector worldwide similar and/or different?
5. Which similarities and differences in the Time Perspectives of people working in the water sector worldwide are likely to be particularly important for water planning and setting research agendas?

Question 4 is central to the empirical study and a nuanced answer is given in Chapter 4, which is further developed theoretically in Chapter 5. In essence, the Time Perspectives of people working in different professions throughout the water sector worldwide were relatively similar as regards how the Structure of time is conceptualized and also the Orientation that people had towards or against the past, present, and future. Significant differences were found, however, concerning the Temporal Extent and the Certainty that people attribute to the objects that motivate their work related decisions and actions (Chapter 4). To briefly answer Question 5, the differences in terms of the Temporal Extent and the Certainty attributed to Motivational Objects were greater across professions than across national borders. This presents interesting consequences for water planning and the setting of research agendas. This answer is expanded and explained in the following section on implications and applications which deals with the sixth and final question:

6. Once similarities and differences in time perspectives are acknowledged and understood, how can this understanding be used to improve water planning and the setting of research agendas at different scales?

The research project did not include empirical tests of implications and applications for water planning and setting water research agendas. To answer the sixth research question potential implications and applications of the empirical conclusions are examined in the next section. The consideration of implications and applications both rested on and raised various theoretical issues, which are reviewed in a theoretical discussion. The Chapter concludes with a list of recommendations.

6.2 IMPLICATIONS AND APPLICATIONS

The conclusions of this research project are potentially quite widely applicable because the domain under investigation is essentially human. For example, it is imaginable that specialists in environmental law could use the quantitative dimensions of the Foreseeable Future, namely Temporal Extension and Certainty, to define culturally specific thresholds beyond which a supplementary dose of precaution may be required. There is currently no general definition of the Precautionary Principle for European Law because its application is context specific and depends on the level of risk that a society considers acceptable (Christoforou, 2003). But it has also been argued that scientists need to translate the Precautionary Principle into more instrumental tools for proactively protecting ecosystems, rather than just anticipating levels of damage (e.g. Tickner, 2003), and the Foreseeable Future could be useful for this purpose.

The Foreseeable Future could also find applications as a 'rough' sustainability indicator. The activities of certain actors might be signalled as potentially unsustainable if a mismatch is measured between the duration of the consequences of their activities and the Temporal Extension of their Motivational Space. Similarly, the quantitative dimensions of the Foreseeable Future could be used by economists challenged by cost-benefit analyses to inform and justify decisions and actions. Cost-benefit calculations rely on inter-temporal choice and temporal discounting (e.g. Frederick et al., 2002; Hardisty et al., 2013).

For these practices, the conclusions of this research project definitely raise relevant questions that tie back to the principles of precaution. But rather than closing this book with a collection of broad implications and applications, the focus is firmly fixed on water planning and setting water research agendas. Hopefully people working on other activities also consider the conclusions interesting enough to apply elsewhere.

The consideration of implications and applications is divided into two sections based on the spatial scale. The first section includes all of the international ambiguity in Motivational Space, with the similarities and differences across all National Cultures and Professional Roles in the water sector. The second section focusses on the ambiguity in Motivational Space across Professional Roles, and then on a subnational scale for different types of countries. Some rather instrumental interpretations of the conclusions are explored in these sections, which are meant to add something to their utility but not as a substitute for the value of the insights themselves. The theories upon which these interpretations rest are described in more depth in the theoretical discussion that follows this section.

6.2.1 International implications and applications

The Time Perspectives of people engaged in the water sector worldwide were compared across Professional Roles and across National Cultures. The implications and applications of their similarities and differences across National Cultures are initially discussed. This discussion is specified for four types of National Culture, typified by the Temporal Extent and Certainty attributed to the objects that motivate decisions and actions (Figure 6.1). A detailed version of this map is explained in Chapter 5. The Type 1 National Culture is most widespread and it includes countries such as the Netherlands, Great Britain, the United States, and Australia, but also Ghana and India. People in these countries tend to be motivated by relatively short-term objects and are also less certain that the goals and events they foresee will actually occur in the future. Type 2 is characterized by a classic 5-year horizon but people in these countries, for instance Germany, tend to be more certain about the objects upon which they base decisions and actions. Countries such as Brazil and Japan are included in Type 3; they are characterized as being relatively certain about long term objects. Type 4 includes countries such as China, where decisions and actions tend to be motivated by relatively long-term objects whereas their level of certainty is relatively low.

As described in conclusions, the differences within countries and across Professional Roles are generally greater than those across National Cultures. In addition, the international similarities and differences in terms of Temporal Extent were found to be large and distinctive within the broader context of Cultural Proximity, whereas diversity in terms of Certainty is less pronounced (Chapter 5). It is important to consider any possible conclusions with this in mind. The differences across countries amplified those found across roles. More specifically, the differences across Professional Roles are likely to be most extreme in Type 3 and Type 4 countries. This is why the Motivational Space covered by these two Types is relatively broad (Figure 6.1). It is interesting to consider this pattern in the light of emerging consensus on two fronts: (1) that water planning and decision making should be participatory (e.g. Webler and Renn, 1995; Pahl-Wostl et al., 2008), and (2) that sustainability challenges require transdisciplinary knowledge production involving actors from outside academia (e.g. Lang et al., 2012). These two developments, which are described in the introduction (Chapter 1), form a context for discussing the implications of the conclusions.

The gap between science and the rest of society

The conclusions may be interpreted to imply that participatory planning and transdisciplinary knowledge production are likely to be more difficult in Type 3 and Type 4 countries because of the relatively large differences in Motivational Space across Professional Roles. In these countries the managers of water

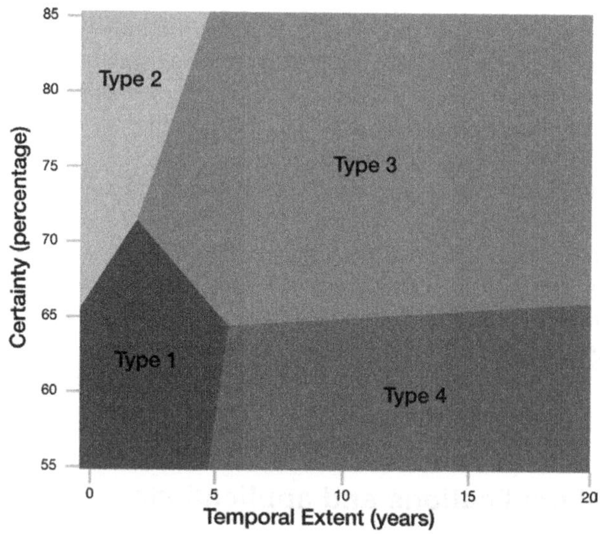

Figure 6.1 Simplified map of ambiguity in Motivational Space across National Cultures.

supply companies, for example, are likely to perceive different problems and be motivated by quite different objects than academics from research institutes. It seems logical to assume that this would make it less likely for people from these two groups to achieve consensus about what the most important problems are, let alone agree on a plan or a research agenda. It would be interesting to investigate whether there are fewer examples of collaborative research teams in Type 3 and Type 4 countries than in Type 1 and Type 2 countries. The quantitative conclusions can be seen to suggest that this is likely to be the case. And the experience of doing the research in the Type 3 countries left the impression that this may be true for Japan and Brazil. In these countries the scientific community for water research seemed to be more abstracted from actors outside academia than in Ghana and the Netherlands. But further research would be needed to substantiate this first impression. Following a similar line of thought, it would be fascinating to investigate whether planning processes are more or less participatory in Type 3 and Type 4 countries than in Type 1 and Type 2 countries. The empirical findings show that motivational objects differ significantly across Professional Roles in these countries, which is likely to make collective planning difficult.

Consensus on problems versus diverse perspectives for change

Heterogeneity in viewpoints can be problematic for planning processes (Van ‘t Klooster, 2007; Hulme, 2009; Muro and Jeffrey, 2012). But diversity in perspectives that across Professional Roles, for example, can also be seen positively as potential for learning, change, and innovation. Ambiguity and even confusion may actually be required to reframe an issue in a way that makes change possible (Dewulf et al., 2005). The Temporal Extent of the objects that motivate Scientists and Managers in Type 1 countries tended to me more similar and short-term than those in Type 3 countries. This may represent a trade-off between (1) potential for consensus across Professional Roles about which problems deserve attention, and (2) the diversity of perspectives that is needed to generate suitable responses to these problems. In addition, scientists working in Type 3 and Type 4 countries may have a greater mandate and capacity to explore temporally distant futures.

It would be interesting to investigate if this has longterm implications. There is a significant correlation between Hofstede's Long-Term Orientation index (LTO) and the increase in Gross National Income between 1995 and 2005 for eighty-four countries (Hofstede et al., 2010). The authors also found a positive relationship between LTO and the growth in the number of patents between 1999 and 2011 for Australia; Brazil; China; Germany; India; Japan; Korea; Netherlands; New Zealand; Norway; Sweden; Taiwan; United Kingdom, and the United States of America. It is beyond the scope of this research project, but a closer look at research and development indicators worldwide could reveal some more remarkable correlations with the Temporal Extent of a country's Motivational Space. As Tainter (1990) described in The Collapse of Complex Societies: "the major factor in understanding these episodes is not that a resource was depleted, but that the respective societies responded in different ways.".

Implications for transboundary cooperation

To highlight some more specific implications of the empirical findings on the international level it would be most interesting to identify sets of countries that share water resources and also have quite different types of Motivational Space. But interestingly enough there are no countries that have been characterised as Type 1 or Type 2 and which share borders with Type 3 or Type 4 countries. The pair of countries that is most geographically proximate and theoretically most different in terms of Motivational Space is Thailand (Type 3) and Bangladesh (Type 1). In considering Motivational Space, however, closer inspection reveals that these countries are on immediately opposite sides of the line between Type 1 and Type 3 (Figure 5.6). They are also separated by, among other things, the Arakan Mountain Range in Myanmar and thus do not share water resources. The potential for pointing out possible implications of the empirical findings for transboundary water planning is thus limited by the data available. An interesting case for future research would be to investigate how Turkey, Syria, and Iraq deal with the Euphrates and Tigris Rivers. This case has unique potential for cross-cultural differences between Europe and the Middle-East regarding Motivational Space (Hofstede et al., 2010).

Clear implications for transboundary water planning may be difficult to establish based on the empirical findings, but it is still interesting to discuss the international ambiguity in Motivational Space in relation to global challenges for the water sector. The International Water Association (IWA), for example, works on thematic programmes that are considered to be common to countries worldwide. As described in the introduction (Chapter 1), these issues generally concern wicked problems and are likely to be framed differently by people from different National Cultures and Professional Roles. Prominent examples include Climate Change and the integrated design of cities. It would be interesting to analyse the membership of the IWA communities and specialist groups that are working on these issues. The conclusions suggest there is likely to be greater cooperation between people from countries that share a type of Motivational Space because they would be motivated by more similar goals and also frame issues more similarly.

It would be interesting to test this hypothesis with empirical research, although it would also be quite remarkable if there were actually distinct networks of people in the water sector worldwide that can be grouped according to types of Motivational Space. Activities that involve people from different types of countries, be they for water planning or setting research agendas, are most definitely influenced by various other cultural dimensions. Language barriers are an obvious example. Communication and cooperation is understandably easier if people share a language. There is also the question of spatial proximity. These are reasons why IWA has set up regional hubs for coordinating knowledge exchange. According to the website of the European Commission, the cost of language services in all EU institutions amounts to about one billion euros per year. With such strong barriers and gravitational forces forming networks of people it is likely to be difficult to distinguish the influence of ambiguity in Motivational Space. But if such groups are discernible, the conclusions would suggest that the networks would initially be formed around Professional Roles followed by groupings around National Cultures.

These ideas may be particularly relevant for the United Nations frameworks for action, such as the Millennium Development Goals (including the post2015 Development Agenda) and the World Water Assessment Programme (WWAP). In addition to the fact that different countries started out with incomparable situations, perhaps the uneven rates of purposeful change across countries can be explained to some degree by differences in Motivational Space. These UN frameworks also rely heavily on processes that are directed at achieving international consensus, for example on multilateral trade agreements, which might be better understood in the light of international ambiguity in Motivational Space. It seems reasonable to assume that an improved understanding of what motivates the decisions and actions of different groups would enable them to find common ground and discuss differences. On the other hand, perhaps such things actually need to be nebulous to conceal ideological differences and make it possible to blindly intertwine different interests to build coalitions (Mosse, 2004). The significance of international ambiguity in Motivational Space for the potential of different groups to affect purposeful change is explored further in the theoretical discussion about windows of opportunity and vulnerability (Chapter 6). Even though the inter-role differences are leading, it is also interesting to briefly explore some international implications for water research agendas before going on to discuss the sub-national scale. For this purpose the most pertinent case is the European Framework Programme for Research and Technological Development.

Implications for European research

The European Commission is currently preparing the next Framework Programme, Horizon 2020, which will shape the future of European research starting in 2014. Most European countries are categorized as Type 1 or Type 2 (Figure 6.1), so it is interesting to look for any traces of Time Perspective in the programme documents. The most obvious potential sign is the 6-year timeframe (2014-2020), which corresponds exactly with the Temporal Extent of Type 1 and Type 2 countries. The duration of the research projects themselves also tends to range between 1 and 5 years. Horizon 2020 represents a further step in the conflation of research and innovation processes; pushing all activities closer to the market. The tendency to emphasize piloting and demonstration projects and support public procurement and market uptake is already evident in the European Innovation Partnership on Water (EIP Water). This approach to EU research and innovation focusses on action groups that are meant to bring together diverse partners from research, industry, finance, and NGOs. Innovation theories such as the Triple Helix model (Etzkowitz, 2012) are at the source of this idea. The conclusions suggest this concept is most likely to work in Type 1 and Type 2 countries where the differences in Motivational Space across Professional Roles are relatively small. This seems promising for Horizon 2020, at least as regards the potential for achieving consensus about the problems. Then again, a similar approach is also being implemented in Type 3 countries such as Brazil (Etzkowitz et al., 2004). Considering the conclusions about international ambiguity in Motivational Space, it would be interesting to compare the performance of these systems for supporting and steering research and development across different types of National Culture. It would also be fascinating to investigate how involving industry and financial partners in setting research agendas influences the time horizon of the projects.

In this section some implications and applications of National level ambiguity in Motivational Space were highlighted for water planning and setting research agendas, which are summarized in Box 6.1. In addition to these conclusions it is evident that an empirical study is needed to explore how Motivational Space may be expressed in the actual activities of different groups. Such a study is, however, beyond the scope of this research project. In the next section the implications and applications of ambiguity in Motivational Space are discussed for the sub-national scale.

6.2.2 Subnational implications and applications

The ambiguity in Motivational Space across Professional Roles is generally greater than that across National Cultures (Chapter 5). This inter-role ambiguity was found to be much more pronounced in Japan and Brazil than it was in the Netherlands and Ghana. The international implications and applications are discussed in the previous section. This section focusses on the subnational scale, but further general observations are also made. A detailed statistical analysis of the similarities and differences across all subgroups is given in Chapter 4, in which Figure 6.3 shows the specific distributions for Temporal Extent. In this section the discussion of implications and applications is simplified by combining the National Cultures and comparing the ambiguity in Motivational Space across Professional Roles (Figure 6.2).

BOX 6.1 Implications of national level ambiguity in Motivational Space for water planning and setting research agendas.

Type 1 - Implications for Water Planning

Adaptive planning may be an appealing concept for Managers in Type 1 countries because there a tendency already exists to accept some uncertainty and focus on short-term goals (2 years). The risk of applying this planning philosophy in these countries is that the planning becomes reactive rather than proactive. This is problematic if the plan involves inflexible infrastructure with a long service life (>10 years). Adaptive planning assumes a vision and a strategy; that is, something to adapt. To deal with the risk of becoming reactive it may be especially important to invest in: (1) the most primary foundations of a plan – the identity, mission, and vision of the organisation, and (2) horizon scanning and futures research to signal potential opportunities and threats that emerge beyond the Temporal Extent of Managers (>2 years). To be included in decisions and actions, however, these signals may need to be translated back in to consequences on the 2-5 year horizon.

Type 1 - Implications for setting research agendas

Managers, Practical Workers, and Scientists in Type 1 countries tend to have a relatively similar Motivational Space. This could enable transdisciplinary knowledge production by making it easier to achieve consensus for setting research agendas. It might also be easier to demonstrate the importance of research for society in these countries. This advantage comes with the risk of limiting all research to topics that have short-term (1-4 year), practical or financial benefits. The growing tendency to include industry and financial partners in agenda setting could amplify this risk (Rappaport, 2005). This could jeopardize the competitiveness of the science and innovation systems over the longer term (several decades). To avert this risk it is common for some scientists to be supported in exploring longer term horizons (>10 years).

Type 2 - Implications for Water Planning

Adaptive planning is likely to be most difficult to implement in Type 2 countries. There is a strong tendency to focus on short-term goals (1-5 years) with relatively certain outcomes. The 'command and control' paradigm is likely to dominate, with a focus on robustness. Scientists, who have the longest Temporal Extent, have an important role in Water Planning; to signal potential opportunities and threats on a longer term (5-10 years). This role could come with problems related to scaling (e.g. global climate model output) down to the level of precision required by decision makers, but this precision does nothing for the accuracy or the certainty of the information (Wilby and Wigley, 1997). To avert the risk of missing windows of opportunity for change it may be wise to set up planning systems that stimulate thinking about longer term goals and events (5-10 years). A 10-year plan documented in a report is unlikely to influence the decisions and actions of managers and practical workers.

(Continued)

BOX 6.1 Implications of national level ambiguity in Motivational Space for water planning and setting research agendas (*Continued*).

Type 2 - Implications for setting research agendas

On an international scale most of the ambiguity in Motivational Space is due to differences in terms of Temporal Extent. The difference between Types 1 and 2 is, however, only in terms of Certainty. People from Type 2 countries tend to be relatively certain that the goals and events they foresee will actually occur. They are also inclined to require more certainty about the future objects upon which they base their decisions and actions. The implications for Type 1 thus also apply for Type 2, but then with the added risk, when it comes to wicked problems, of wanting to calculate the unknowable. Besides being limited in Temporal Extent, there is the risk that the research agendas will tend to focus on questions for which the outcomes are known in advance. To avert this risk it is common for some scientists to be supported in doing fundamental, explorative research.

Type 3 - Implications for Water Planning

Managers in Type 3 countries tend to be inspired by the classic 5-year goals and are relatively certain about that which they foresee. Scientists in these countries tend to think about plans for the coming generations (10-20 years) and are rather certain about their plans, which means that they might run the risk of fashioning utopian visions. On the other hand, this great capacity to be motivated by temporally distant objects might be employed by Managers during the visioning stages of Water Planning to explore pioneering alternatives. Adaptive planning could seem unnecessary to people who are rather certain about the long term, and robustness may be considered a greater goal. The risk associated with this approach is that the situation changes in ways that have not been envisaged while major investments have been made in a certain view of the future, which may result in redundant infrastructure and higher costs. To avert this risk an actor can invest in flexibility and reversibility.

Type 3 - Implications for setting research agendas

Scientists in Type 3 countries tend to be motivated by goals and problems on a much more distant time horizon (10-20 years) than Managers (4-6 years) and Practical Workers (2-3 years) while also being relatively certain about what they foresee. The interdisciplinary diversity could provide great potential for innovation but may also hamper transdisciplinary knowledge production. There is potential for researching more long-term questions with the risk of drifting away from what is practical and relevant to society. It could be difficult, but important e.g. for receiving finance, to translate research findings back to implications for Managers on a 4-5 year time horizon. Type 3 countries are also characterized by the tendency to want to know the unknowable. To avoid only starting studies for which the outcomes are foreseen it is common to include some fundamental, explorative questions in the research agenda.

Type 4 - Implications for Water Planning

People making plans in Type 4 countries tend to have longterm visions (5-20 years) and require less certainty than those in other countries to make decisions and actions. These characteristics may make it possible to plan for decennia into the future. As with Type 3 countries, however, the risk is that long-term investments are made in a future that doesn't eventuate. On the other hand, planners in these countries can more easily accept uncertainty as part of the nature of the future and take it into account. The diversity in Motivational Space across Professional Roles could make collaborative planning difficult but it also brings great potential for adaptive planning because people performing different roles can cover the competencies needed to practically implement this approach (e.g. Medema et al., 2008).

Type 4 - Implications for setting research agendas

The difference between Types 3 and 4 is primarily in terms of Certainty. The implications for Type 1 thus also apply for Type 2, except for the need make room for fundamental, explorative questions in the research agenda. The combination of low certainty and a long-term horizon (10-20 years) could amplify the risk of the science system becoming isolated from society. On the other hand, if transdisciplinary knowledge production is managed well, taking the ambiguity in Motivational Space into consideration, there may be great potential in the diversity of views for innovation.

(*Continued*)

BOX 6.1 Implications of national level ambiguity in Motivational Space for water planning and setting research agendas (*Continued*).

Types lumped - Implications for Water Planning

Ambiguity in Motivational Space is likely to be one reason why international initiatives, like the UN frameworks, rest on the 'Nirvana' concepts of Integrated, Participatory, and Adaptive water planning. There may be general consensus that the world consists of dynamic systems including matter and people that need to be harmonized, but more specific plans for the future are basically context specific. And the conclusions show that even within a given context, the planning problems may be framed differently by people fulfilling different Professional Roles. Perhaps this diversity is part of what makes the human species resilient to change (Folke et al., 2002). But what are the implications for international collaboration in response to wicked problems such as Climate Change and biodiversity loss? In an ideal situation we might dare to dream like the Brazilians, embrace some guiding principles like the Japanese, be pragmatic like the Dutch, and forgiving like the Ghanaians. The aim of this research project was to reveal the ambiguity in Motivational Space and the main recommendation is to frame problems and plans consciously with this ambiguity in mind.

Types lumped - Implications for setting research agendas

The cross-cultural differences in Motivational Space could complicate the setting of international research agendas, such as that of the International Water Management Institute (IWMI). If only scientists were involved in agenda setting then the Temporal Extent of Motivational Objects would still vary across individuals between 4 and 20 years. Prioritizing problems for the next 4 years is likely to result in quite a different research agenda than if that horizon was 20 years. Agenda setting for transdisciplinary knowledge production, which includes people from different Professional Roles, will include even greater differences. It may be intelligent for the science systems in Type 1 and 2 countries to work together with Scientists from Type 3 and 4 countries towards setting longer term agendas for fundamental research. When it comes to transdisciplinary research it is important to match the goals of the research programme with Motivational Space of the agenda setters. Ambiguity in Motivational Space, once acknowledged and understood, can also be included by design in research projects to facilitate co-learning and generate new perspectives.

The area of the Motivational Space in Figure 6.2 is defined using the interquartile ranges of Temporal Extent and Certainty for each Professional Role. This means that 50% of all of the individuals in each of these groups is typically motivated by objects that fit into the indicated Motivational Space per group. The upper and lower quartiles have been trimmed. Temporal Extension differed significantly across all Professional Roles in every National Culture except between Practical Workers and Managers in The Netherlands. On the other hand, Certainty did not differ significantly between Scientists and Practical Workers, and only in Japan did Certainty differ significantly across all Professional Roles (Chapter 4). Figure 6.2 needs to be understood with these national nuances in mind. Figure 5.7 in Chapter 5 is a more detailed map showing the median of all twelve subgroups. The map below is meant to epitomize that inter-role ambiguity as a simple basis for the discussion.

Ambiguity in Motivational Space across Professional Roles obviously has various implications depending on the national and regional context. The Dutch context is explored in more depth to highlight several examples of specific consequences and then some more general conclusions are summarized in Box 6.2. The Dutch case was chosen because of its familiarity. This country is also categorized as Type 1, which includes most of the countries for which data is available Great Britain, the United States, Australia, Ghana and India. Some of the implications and applications may thus be more broadly applicable. But before descending to the subnational scale it is important to briefly reflect on how European systems influence Dutch water planning and research agendas. As a European member state the Netherlands is committed to the European Water Framework Directive (WFD 2000/60/EC of the

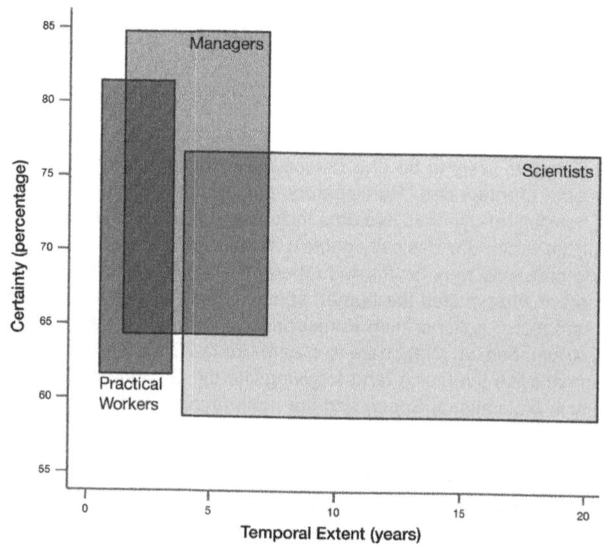

Figure 6.2 Simplified map of ambiguity in Motivational Space across Professional Roles (Space defined by interquartile range: 25% trimmed).

European Parliament and of the Council), which aims to promote sustainable water use based on a long-term protection of available water resources. This WFD stipulates objectives with time horizons varying between 6 and 15 years after the date when it entered into force (23 October 2000). These objectives combine with other European agreements, such as the Habitats Directive and the Birds Directive, to form a legislative setting within which the national scale plans are made. From the perspective of Temporal Extent it is interesting to note that the European Parliament is elected every 5 years, while many of the mandatory reports, such as the WFD river basin management plans, are updated every 6 years.

Terms in government

It is interesting to view these time horizons within the context of the terms served by government. The principle of turnover is fundamental to theories about democracy and can be traced back to ancient times (Struble, 1979). Incidentally, annual turnover was most prevalent in the ancient systems in Italy and Greece. The mode in Europe is now to follow 4 or 5 year cycles, which also seems to be reflected in the local plans. The maximum term of the Dutch House of Representatives was set at 4 years in 1848. Since then the actual period of government has been fluctuating between 1 and 4 years, with re-elections being called for various reasons such as a collapse of the governing coalition. It would be interesting to investigate the rationale behind this time horizon and also its effects on planning at different scales. In any case, Figure 6.2 indicates that this is also the average temporal extent of the objects that motivate the decisions made and actions taken by Managers. On a national scale the maximum term of government in European countries, including the Netherlands, also matches that of the Temporal Extent of their Motivational Space (Figure 6.1). The conclusions are inconclusive when it comes to determining whether the Motivational Space of the individuals determined this time horizon, or the other way around. In any case, the scientists seem to be remarkable insofar as objects beyond this time horizon motivate them. In the light of this conclusion it would be particularly interesting to augment the empirical study by interviewing ‘policy makers’ as a fourth Professional Role.

The temporal cycles of national and European government seem to be reflected in the water plans and research on more local scales. The Dutch have a National Policy Strategy for Infrastructure and Spatial

Planning that has a time horizon of 20-30 years. And in 2009 the cabinet of the Netherlands defined a long-term vision (2050-2100) for water management. At first glance these documents might suggest that the Temporal Extent of the Motivational Space perceived by policy makers is longer than that of the three Professional Roles included in this study. On the other hand, policy makers may have mutual expectations of one another and also feel pressure from society to take longer term change into consideration because of their perceived role in society. It would be fascinating to investigate from where this social pressure originates, because the Temporal Extent of the people interviewed in this research project was considerably shorter. An empirical study is also needed to assess whether the collective intelligence of groups (e.g. Woolley et al., 2010), or other in-group and out-group interactions involving normative pressure to comply to ideals (e.g. Abrams et al., 1990), enables groups to have a longer term view than that of the individuals they comprise. The challenge for such a study would be to dissect the rhetoric and determine which motivational objects are causally efficacious.

Planning on different levels of abstraction in the Dutch water sector

Although it was not intentionally studied, the experience of interviewing people in all four countries supports the idea that long-term thinking is generally considered to be in some way superior to short-term thinking. After being interviewed people tended to confirm the results while also expressing some embarrassment about their 'limited' Temporal Extension. There may be various reasons for this, but the worldwide prevalence of prophets, fortune tellers, and tea-leaf readers throughout history seems to suggest that any promise of taming prospective uncertainty is likely to be prized and revered. But is the 2050-2100 time horizon mentioned in the long-term vision for water management in the Netherlands an indication of temporally extensive Motivational Space? Such hazy time horizons are commonly chosen for the practical reason of extending the discussion beyond the Motivational Space in which actors have vested interests so as to enable them to consider alternatives without getting bogged down in advocacy. Consistent with this idea, the long-term vision is more exploratory than binding. In contrast, the National Water Plan that also came into effect in 2009, and which unlike the long-term vision has practical and legislative consequences, is set to a time horizon of 6 years (2009-2015). Officially, Dutch integrated water policy and management is reviewed following a 6 year planning cycle (Arnold et al., 2011). To summarize, the hierarchy of planning processes for the public water supply in the Netherlands is presented in Table 6.1.

If the discussion of implications and applications is restricted to the groups of people for whom empirical data is available then the Water supply plans, Multi-year investment plans, and Business plans of the 10 Dutch water supply companies are of particular interest (Table 6.2). The Time horizons of these plans vary between 5 and 30 years and are primarily driven by prognosis of water demand. These prognoses are the basis for the Water supply plan and the Multi-year investment plan (Table 6.2). It is interesting to note that prior to 1990 the prognoses of domestic water demand in the Netherlands was calculated based on statistical extrapolation of growth in the past. This approach worked fairly well as the demand increased quite constantly from 1950 until 1990, when it peaked and levelled off (Baggelaar and Geudens, 2011). Planners explained this peak in retrospect, by highlighting the market saturation of bathroom showers and washing machines together with an increase in the efficiency of appliances. Only after this peak, however, were bandwidths of uncertainty included in the prognoses based on the population projections of the national bureau of statistics (CBS). This goes to show that a 10-year prognosis does not necessarily entail Temporal Extension of 10 years. The mode for the Temporal Extent of Dutch (3 years) is more likely to be reflected in the fact that these prognoses are regularly updated at 3-5 year intervals (2002, 2005, 2008). This could indicate an expectation to foresee 10 years into the future whereas those making the prognosis feel more comfortable with the 3-5 year horizon. In any case, this explanation would be supported by the conclusions. The discussions to this point suggest that the Motivational Space of the Dutch is reflected in their planning cycles. It is now interesting to look for some more concrete signs of this Temporal Extent before reflecting on the ambiguity across roles.

Table 6.1 Hierarchy Scale Name of Plan of planning processes.

Scale	Name of Plan	Time horizon (years)	Organisation
National	Policy plan for drinking water and industrial water supply	20-30	Ministry of Infrastructure and the Environment (LenM)
	National Water Plan	6	
National Section	Ten-year prognosis of water demand (3-5 year updates)	5-10	Association of Dutch Water Companies (VEWIN)
Provincial	Provincial policy plan	5-10	Provincial Government
	Provincial management plan	5-10	
Regional	Regional plan	5-10	
Municipal	Land use plan	5-10	Municipal Government
Company	Water supply plan	10-30	Water Supply Company
	Multi-year investment plan	5-10	
	Business plan	5	

Asset Management

As with most established utility services worldwide, the infrastructural assets managed by the Dutch water supply companies are vast and valuable. The transport and mains supply network, for example, includes more than 100 thousand kilometres of pipes and the tangible fixed assets are estimated to be worth about 5.5 billion Euros (Geudens, 2010). The service life of this infrastructure can be estimated from an economical or a technical perspective. Traditional asset management takes the technical service life as leading. This largely means that plans for replacement are based on the point in time at which the component is worn out and no longer satisfies the function for which it was designed. The economical service life is generally shorter than the technical service life because the operating and maintenance costs and risk of failure increase as an asset ages, which makes replacement more economical at a certain point. There is a trade-off: replacing infrastructure too soon results in loss of capital, while too long a delay can lead to failure and increased costs. To give an indication of the time horizons in question the average technical service life of different types of water supply assets is summarized in Table 6.2.

According to De Graaf and Der Brugge (2010) innovation is limited and maintenance and optimization dominate in the Dutch water sector because of the temporal mismatch between the planning cycle (4-6 years) and the technical lifespan of infrastructure (40-60 years). As a response De Graaf and Der Brugge suggest a transformative approach to asset management, which involves using infrastructure that has a shorter lifespan and is thus more adaptable to changing circumstances. On the other hand, the costs associated with removing and installing infrastructure may be far greater than the cost of the infrastructure itself. Consider, for example, the costs of replacing pipelines that run under highways or cities. It could be most cost-effective to postpone the replacement of existing pipelines for as long as possible and invest in better methods of monitoring their condition (Beuken and Mesman, 2011). Managers and practical workers who were found to be motivated by objects on a median time horizon of 2 years are laying new PVC pipelines that have a technical service life of several centuries. But since the pipes themselves are relatively inexpensive this may not be particularly problematic. Real issues arise when the resources are costly and the interventions are irreversible. Following this line of thought, one potentially problematic activity is the exploitation of groundwater reserves. This idea is developed further in the theoretical discussion on Tipping Points and Institutional Irresponsibility because such questions merit more nuanced treatment and conceptual underpinning.

Table 6.2 Average technical service life of different types of assets (adapted from www. kostenstandaard.nl/, retrieved October 2013).

Type of Asset	Indicative Technical Service Life (years)
Water intake/abstraction	40
Civil engineering/buildings	40
Mechanical components	25
Electrical components	20
Process automation facilities	10
Transport pipelines	60
UF Membranes/HF	8
Membranes NF/HF	5
Seawater RO Membranes	4
Replacement activated carbon (new)	2
Replacement activated carbon (regeneration)	15
Other (filter sand etc.)	10

Phasing plans for motivation

Further to a better understanding of the existing planning strategies, the phasing of plans can now be tailored to the Motivational Space of specific actors to define milestones most likely to motivate decisions and actions (Figure 6.3). Figure 6.3 shows how the Temporal Extent of two hypothetical groups group of actors is used to break the final objective (what) and deadline (when) into three targets (when) and then milestones (what). Group 1 has a shorter Temporal Extent than Group 2 which means that they need two milestones (A & C) in addition to the final objective to keep them motivated. For Group 2 only one extra

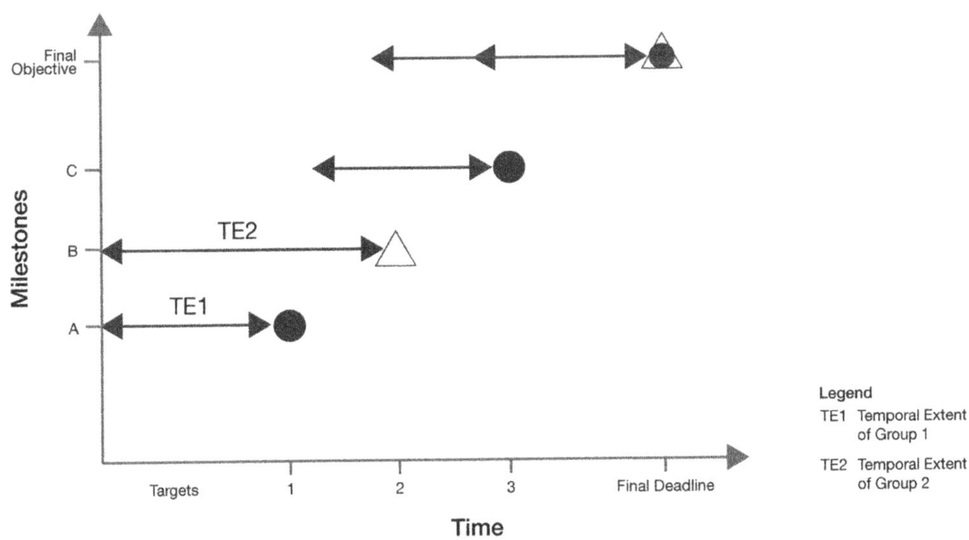

Figure 6.3 Phasing plans according to the Foreseeable Future of the target group.

milestone is needed (B). On the subnational scale, concerning participatory planning, the number of groups obviously depends on the ambiguity in the Motivational Space of those involved. For this simplified model Group 1 could consist of Managers and Group 2 of Scientists.

If the final objective and deadline is known then the Temporal Extent of the various actors can be used to backcast targets and define the related milestones. This is a reversal of the normal approach to scheduling, for example in Gantt charts, and might thus seem counterintuitive. Targets (when) are usually defined depending on the predicted amount of time needed to complete a certain milestone (what) and the dependencies between milestones. Most dependencies are unlikely to change: When building a house it is, for example, generally preferable to (A) lay the foundations before (B) putting up the walls. The novelty of the approach to phasing depicted in Figure 6.3 is that, after the final objective and deadline have been defined, targets based on the Temporal Extent (when) of the relevant actors are used to define the milestones (what) rather than the reverse. In theory this could lead to greater motivation towards achieving the milestones.

Implications for hierarchical models as analytical devices

Insight into the Foreseeable Future can be seen as a general tool for explicitly addressing multilevel temporalities. The conclusions could be used in this way to improve analytical models of social behaviour that are dynamic and based on hierarchical reasoning. Genus and Coles (2008) provide a clear overview of how hierarchical reasoning, using Multi-level frameworks for example, is increasingly used to analyse long-term socio-technical transitions. These analytical models are designed to represent the complexity of multi-level interactions dynamically across different timescales (Geels and Schot, 2007). Such models can be used beyond their applicability by transition managers who overgeneralize and aim to reshape entire trajectories of technological innovation (Berkhout et al., 2004). But in their function as analytical devices, including the Motivational Space of relevant groups of actors may augment them. This could improve the theories by adding an empirically derived basis for defining the temporal scales of some social processes. The conclusions about differences in Temporal Extension across Professional Roles are likely to be especially interesting for this application. This idea is developed further in the theoretical discussion on change in complex socio-ecological systems.

Implications for Horizon Scanning

The conclusions about differences in Temporal Extension across Professional Roles are also useful for directing horizon scanning activities; that is, the collection of strategic intelligence information. While working towards a milestone, the Temporal Extent of a group's Foreseeable Future can be used to estimate the horizon beyond which emerging opportunities and threats are unlikely to be perceived or identified. This horizon thus also represents the point beyond which horizon scanners need to focus their efforts. And once potential opportunities and threats have been identified the same Temporal Extent is very useful for helping the horizon scanners to translate their findings into indicators upon which managers are more likely to act. The applications for strategic and tactical planning could be quite meaningful. Scientists might be explicitly employed to scan the temporal horizon for impeding threats and opportunities. The main challenge will involve reframing the insights as motivational objects for the decision makers (Figure 6.4). This idea is discussed in more detail in the theoretical discussion on windows of opportunity.

Implications for setting research agendas

For setting water research agendas, a similar application to that of water planning is conceivable. Once the Temporal Extent of the Motivational Space perceived by the agenda setters is known, then it may be

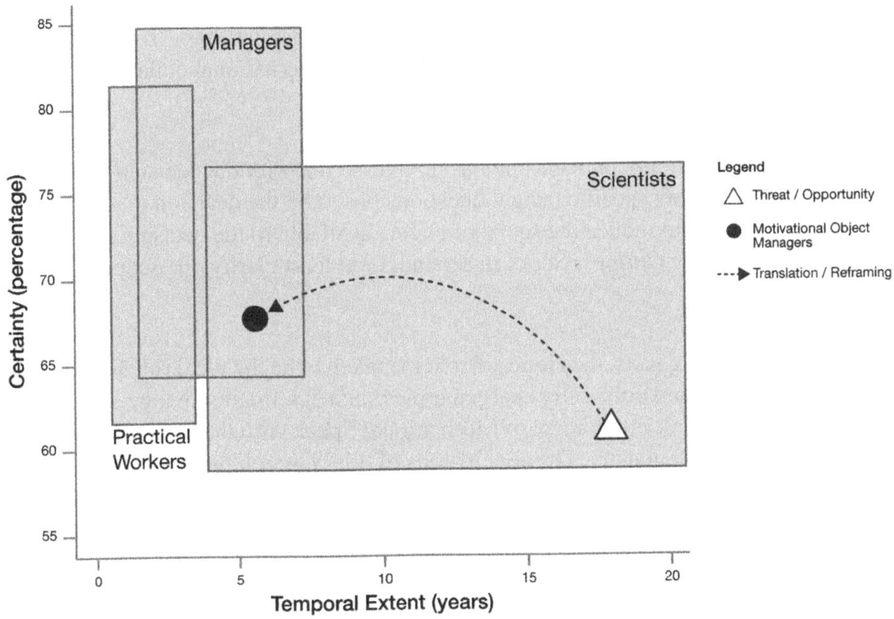

Figure 6.4: Roles for signalling opportunities and threats taking motivational space into account.

possible to predict the type of research that will be programmed. Looking at the ambiguity in Motivational Space across Professional Roles it seems unlikely that Practical Workers would prioritize research about, for example, advanced synthetic biotechnologies that may pose opportunities and threats twenty years from now. Allowing a homogenous group of fundamental scientists to prioritize research questions is also unlikely to result in a research agenda that is focussed on, for example, market ready measures for optimizing the management of conventional sand filters. With that in mind it would be interesting to augment the conclusions of this project with an empirical study into what have been the main drivers of change in the focus of scientific research for the Dutch water sector in the past? E.g.: political changes; group dynamics; ideologies; environmental circumstances, and the Motivational Space of the agenda setters.

For transdisciplinary knowledge production the composition of the team of agenda setters as regards their Professional Roles can be purposefully selected to generate different types of research. On both the national and the international scale the conclusions suggest that it might be unwise to make inter-role diversity a requirement for all research (funding). This would probably result in a lot resources being spent on the (re)framing of problems. Then again, Rittel and Webber (1973) argued that the process of defining a wicked problem is the same as the process of solving it. In any case, the purpose of a research programme must be reflected in the Motivational Space of the agenda setters who determine which topics are to be researched. And if transdisciplinary research is preferred then the conclusions suggest that substantially more time and finance needs to be dedicated to framing the problems than is usual for more conventional research projects. Assuming that the result of transdisciplinary framing is to choose topics that are located where overlapping occurs between the Motivational Spaces of people from different Professional Roles, the conclusions indicate that the most probable Temporal Extent will be about 4 years. In Type 1 countries this is horizon is likely to be further shortened to 2 years. These time horizons do not match those intrinsic in the consequences of current human activities, let alone the lag times associated with global change such as that occurring in the climate system. These issues are discussed further in the theoretical discussion on tipping points and institutional irresponsibility.

Looking at Figure 6.4, the conclusions may also be seen to suggest that scientists have an important role to play when it comes to extending the Foreseeable Future. Scientists have the longest Extension, the highest in-group variance, and are also typically accustomed to international collaboration. Pielke (2007) defined 4 roles for scientists:

The Pure Scientist: Focuses only on facts and has no interaction with the decision maker.
The Science Arbiter: Answers specific factual questions posed by the decision maker.
The Issue Advocate: Seeks to reduce the scope of choice available to the decision maker.
The Honest Broker of Policy Options: Seeks to expand, or at least clarify, the scope of choice available to the decision maker.

As the title of Pielke's book suggests, the Honest Broker is taken to be the ideal role for scientists working on transdisciplinary research and politically charged topics such as Climate Change. Assuming that this is the case, it is quite logical to link the concept of Motivational Space with the idea of 'expanding the scope of choice available' to decision makers. The conclusions of this research project show that if the decision maker is a Scientist, it follows that the objects motivating their decisions are likely to span a greater temporal extent than if they were a Manager or Practical Worker. In this sense Scientists may have a role to play in expanding the 'scope of choice' by maximizing the period available for intervening and/ or minimize the time needed to adapt. To fulfil this task effectively Scientists would need to signal potential opportunities and threats that lie beyond the Temporal Extension of decision makers' Motivational Space and translate these insights into information by which the decision makers are motivated. But to remain an Honest Broker the Scientists must also be careful not to conflate the use of predictions as a means to advance science with their use as a means to advance policy (Sarewitza and Pielke, 1999). These ideas are at the heart of the concept of a 'strategic learning alliance', which the author has been working to design and develop since 2009 in the form of Dutch Water Sector Intelligence.

A strategic learning alliance for attending to ambiguity in problem perception

A strategic learning alliance is a form of transdisciplinary futures research that is geared towards providing strategic input for water planning and prioritising research questions for the science system. The Dutch Water Sector Intelligence concept involves bringing scientists, planners, and decision makers from throughout the water sector together several times per year to collectively develop strategic insights that they use as conceptual building blocks for their individual organisations and for groups of partners collectively (Segrave et al., 2011; Frijns et al., 2013). To feed the alliance with insights a group of scientists examines and reports on a wide range of trends throughout each year. The sources of these studies include (online) documents and also undocumented trends that the scientists tap into via networks and conferences. The knowledge acquired through this research is used to choose topics for the transdisciplinary 'think-tank' sessions. These sessions begin with hierarchical learning from experts on the topical trends who fuel the alliance with current, specialist knowledge. But the emphasis is on active, intentional co-learning in smaller groups of varying composition. Debate and (inter)discursivity are used in these groups to provoke reflexivity. Carefully tailored group processes reveal and test participant's assumptions and facilitate generation and analysis of alternatives to ascertain whether or not adaptation is needed for changing circumstances.

Practical experience of the strategic learning alliance concept revealed its main shortcoming to be the limited involvement of the most influential decision-makers. The outcomes of the alliance may thus be restricted and indirect. A proper study is needed to determine why this is the case, but one reason may be the diversity of the group. As described earlier this diversity is both a strength and weakness: one person may be well acquainted with a theory or concept that is new and challenging to another. People also learn at different rates and have different capacities. This is acceptable if the positions are reversed for different subjects, but if the knowledge exchange is overly unidirectional then the benefits are less mutual than is needed to make co-learning inspirational and to maintain the flow between boredom and anxiety (Csikszentmihalyi, 1975). This is just one of the issues that need to be addressed; the development of this alliance is far from complete. But the empirical findings of this research project make the ambiguity in Motivational Space across Professional Roles explicit, which is useful for designing ways of benefitting from this diversity in a transdisciplinary strategic learning alliance. Such an alliance can be very valuable if it expands the scope of choices available to decision makers.

In this section water planning and research has been discussed in relation to the ambiguity that was found in Motivational Space across Professional Roles. The implications and applications are summarized in Box 6.2 and this chapter concludes with an extension of some of the theories that tie in with the interpretations of these conclusions.

BOX 6.2 Implications of ambiguity in Motivational Space across Professional Roles for water planning and setting research agendas.

Practical Worker - Implications for Water Planning

The Motivational Space of Practical Workers is most similar across National Cultures and the subnational ambiguity within this group is also minimal. This could be beneficial if there is a need to develop consensus about the tasks at hand during the later stages of planning. On the other hand, Practical Workers install and operate infrastructural assets that have a service life that extends well beyond the Temporal Extent of their Motivational Space. For participatory planning the role of the Practical Worker may thus be focussed on detailed description of problems concerning the next 2 years and judging the feasibility of response strategies. If Practical Workers are dominant in defining the Motivational Space for planning then there may be a risk of missing opportunities and threats on the 2-20 year time frame.

Practical Worker - Implications for setting research agendas

Practical Workers have an important role to play in setting the agendas of applied research programmes and research and development activities that are geared towards 2 year market-ready tools and concepts. The empirical findings also indicate that groups of practical workers are more likely than people from other Professional Roles to agree on the nature of the problems that need to be addressed by the research. For transdisciplinary research the Practical Workers and Scientists share little Motivational Space so more resources and time needs to be reserved for facilitating co-learning and carefully framing the problems from the outset.

Manager - Implications for Water Planning

For participatory planning Managers, especially in Type 3 and 4 countries, could be important linking pins across levels of abstraction from theory to practice. In Type 1 and 2 countries, however, managers may need longer Temporal Extension to fulfil this role. Managers are generally responsible for choosing between planning options and tend to require more certainty than people fulfilling other roles. Even so, these decisions generally have consequences that extend much further into the future than the Motivational Space of this group. This raises issues about responsibility and it is important for Managers to consider carefully how they legitimize their decisions. Age-old and irreplaceable resources should be avoided. It could also be worthwhile to investigate possibilities for investing in infrastructure that has a shorter lifespan, however this is only likely to be economical for above ground infrastructure for which the installation costs are relatively low.

Manager - Implications for setting research agendas

One purpose for research is to expand the scope of choices available to managers. For this type of research the managers obviously need to be involved in the agenda setting. The empirical findings suggest that the problems that

(*Continued*)

BOX 6.2 Implications of ambiguity in Motivational Space across Professional Roles for water planning and setting research agendas (*Continued*).

Managers are likely to identify will have a Temporal Extent of 4-5 years. If there is an intention to explore more distant time horizons then it is likely to be beneficial to include Scientists in the agenda setting process. For the agendas of transdisciplinary research programmes, the Managers could be important linking pins across levels of abstraction from theory to practice. For fundamental, ground-breaking research it may be difficult to motivate Managers to invest in opportunities, threats, or solutions that are unlikely to materialize within the next 10-20 years. This may be why Scientists sometimes promise 4 year outcomes and then proceed to keep pushing the date forward.

Scientist - Implications for Water Planning

The in-group variance of Scientists, regarding the Temporal Extent of their Motivational Space, is higher than for any other Professional Role, both on the international and national scales. The median Temporal Extend for this group is also the longest, particularly in Type 3 and 4 countries. In addition to the different roles of individual experts as described by Pielke (2007), the science sector might also fulfil an important function in society when it comes to signalling potential opportunities and threats on a 5-20 year time horizon and making contingency plans. For participatory planning Scientists could be specifically employed for this role. The challenge for these Scientists is to translate their insights into objects that motivate the decisions and actions of Managers. The maps of ambiguity in Motivational Space can be used to guide this translation step.

Scientist - Implications for setting research agendas

The in-group variance of Scientists as regards the Temporal Extent of their Motivational Space might be related to differences across disciplines: Microbiologists are likely to view problems on a different scale to geologists. Setting interdisciplinary research agendas is likely to involve quite some ambiguity in Motivational Space, let alone transdisciplinary research that includes Managers and/or Practical Workers. It is thus important for scientists to explicitly define the Motivational Space for all research agendas. And this includes not only the temporal and spatial scope of the systems or problems of interest, but also the Temporal Extent of the researchers. Scientists may also pay more deliberate attention to the role they could play in making societies more sustainable. Considering the nature of change in complex socio-ecological systems (described in the threoretical discussion), maximising the Temporal Extent of the Motivational Space that motivates decisions and actions could be critical.

Roles lumped - Implications for Water Planning

The conclusions showing ambiguity in Motivational Space explain, to some extent, the difficulties experienced in translating wicked water problems, commonly signalled by scientists, into practical response strategies. During the first stages of planning, when problems are framed, the ambiguity in Motivational Space needs to be made explicit so that intelligent choices can be made about the planning horizons. Questions also need to be asked about the Time Perspectives of the people making the plans and that of the consequences of their decisions and actions. Considering the conclusions, it is likely to be beneficial to continuously engage in Horizon Scanning to identify emerging opportunities and threats as soon as possible and expand the scope of choices available. Scientists are likely to be most fit for this role. The maps of Motivational Space can be used to determine the time horizons for scanning and to translate the findings into indicators that decision makers are more likely to act on. Those concerned with the process of participatory planning can use the conclusions about the Foreseeable Future of different groups to stage projects using time periods that keep the different groups motivated.

Roles lumped - Implications for setting research agendas

The ambiguity in Motivational Space across Professional Roles is both a hindrance and a benefit for transdisciplinary knowledge production. Rather than compromising all research to accommodate the Motivational Space of all professional roles, which is likely to limit the problems to those with a time horizon of up to 4-5 years, it is important to match the team with the goal of the research. Considering the amount of time, money, and thought invested in answering research questions there generally seems to be disproportionately little invested in selecting which questions deserve attention. Transdisciplinary knowledge production could bring change in this regard because substantially more effort is needed to (re)frame the problems if people from various Professional Roles are involved. The maps of ambiguity in a Motivational Space can be used to design and facilitate these framing processes. The Foreseeable Future Multi-measure Method has also proven a useful tool that could be used to define the ambiguity in the Motivational Space of those involved in setting any given agenda so that it can be dealt with explicitly.

6.3 THEORETICAL DISCUSSION

The discussion of implications and applications raised some questions that warrant deeper theoretical discussion. This section also includes some possible hypothesis for future research. Interestingly enough, these hypotheses seem to support some popular generalisations, which could either say something about the ideas themselves or perhaps just about the nature of speculative hypotheses. In any case, the first concerns the idea of technological leapfrogging. The notions of fear, hope and progress are also discussed in relation to the Foreseeable Future. In the most fundamental sense, much of the discussion about implications and applications of ambiguity in Motivational Space rests on theories about change in complex socio-ecological systems. These theories are discussed with specific attention given to the concept of 'tipping-points' and some ethical questions concerning the mismatch in the Temporal Extent of Motivational Space relative to that of the consequences of decisions and actions. The discussion ends on a positive note with an exploration of windows of opportunity. This chapter is concluded with a list of recommendations.

6.3.1 Technological leapfrogging: who will leap who?

The Foreseeable Future of Type 3 and 4 countries could enable them to leapfrog Type 1 and 2 countries during the coming decennia. It has been argued that countries need not develop by trailing the technological progression pathways followed in the past and that it is possible to skip obsolete technologies and adopt the current cutting-edge alternatives (e.g. Goldemberg, 1998). The wildfire-like spread of mobile phone technology is commonly cited as evidence supporting this idea (e.g. Stump et al., 2008). But there has also been criticism of the theory of technological leapfrogging related to sustainable development, particularly concerning energy strategies in response to Climate Change. Developments in the automotive industry, for example, once again show how practice can contrast with theory (e.g. Perkins, 2003; Tukker, 2005). Even so, a temporally extensive motivational space could be one of the characteristics that enables technological leapfrogging. The assumption behind this idea is that revolutionary innovations can be more readily recognized and realized if longer-term opportunities and threats are used to motivate decisions and actions. On the other hand, circumstances might just change in such a way that short-term incentives, such as price, encourage people to adopt a new technology without a longer-term vision being necessary at all.

From the perspective of those who are crouching to be leapt, likely candidates may include countries seeming to have reached 'un surplace'; that is, balancing motionless with a conservative, short-term outlook as if frozen in fear of losing what they have. Type 1 and 2 countries that are currently well-off and focussed on maintenance and management could be most at risk. This reflects another stereotypical principle: the handicap of a head start (original Dutch: Wet van de remmende voorsprong). The process of making progress, or adapting to a certain niche, can reduce incentives for attempting further progress and increase resistance to change. Past investments in infrastructure, such as flood defences and distribution pipelines that have improved the quality of life for millions of people, may now inhibit organisations in the water sector in their capacity to change.

Another factor possibly contributing to the handicap of a head start may be the Endowment Effect (Thaler, 1980): For offering up something that they have people generally demand much more than they would be willing to give to acquire that same thing in the first place. If a group of people enjoy all they need, and more, fear of change and restrictive conservatism may ensue. This is a subject that would be exceedingly interesting to investigate further in future research, specifically as regards the nature of the Motivational Objects that different groups of people work towards. Does the Motivational Space of a group shrink as their satisfaction with life increases? How does the Motivational Space of a group or organisation relate to their openness, or hostility, towards innovation and change? Could the Temporal Extent of the Motivational Space of Type 3 and 4 countries enable technological leapfrogging or help them to keep ahead. These are intriguing questions that could be investigated with future research.

6.3.2 Fear, hope and progress

Perhaps fear of regression from the status quo leads to a shorter-term perspective whereas hope for a better state of affairs inspires longer-term thinking. Hope is positively related to individual job performance (Youssef and Luthans, 2007). And the collective emotional orientation of groups can also be gauged in terms of hope and fear, which are characteristics that influence both views and actions (Smith, 1999). Fear, being grounded in the experiential present and based on the memorized past, tends to dominate hope, which involves the more mentally challenging cognitive anticipation of the future (Jarymowicz and Bar-Tal, 2006). Fear is a more primary and automatic emotion, which can work against longer-term progress. Theories about the psychology of risk aversion use related ideas, such as the desire for security versus the desire for potential (e.g. Lopes, 1987).

Around the beginning of 2011, during the Arab Spring, there was a strikingly marked mismatch between the typical questions and opinions being posed by European and American political analysts and journalists and the perspectives given by the protesters. The young Egyptians, Libyans and Tunisians were not interested in steadying the situation - they were striving for change. The protesters were busy breaking through fear barriers and toppling dictators and they just looked confused when the foreign journalists, who were concentrating on questions of stability and security, kept asking about ways to tame and contain the flames.

A lot of frustration and courage was needed to break free from the existing situation and power structures. Experts from the Middle East and North Africa have explained how the revolutions were inspired by incredible optimism and hope, but that to continue the transition a slower, more fundamental, cultural shift is needed (Heideman et al., 2012). If hopes get mixed up with short-term utopian fairy tales then the inevitable disappointment than ensues can allow the fears to creep back in and dominate behaviour. Perhaps a long-term perspective is not needed to start a revolution, but it may be essential to accomplishing enduring, directed change. This idea is developed further in the next section on change and 'attractors' in complex socio-ecological systems.

It would be interesting to divide people's Motivational Objects into hopes, fears, and aspirations (He and Zhou, 2013) and then compare Temporal Extension across categories. To speculate about some implications of this line of thought for water planning: Although both water sectors are highly developed, the Dutch may be at risk of succumbing to the handicap of a head start because they seem to be focussed on protecting what they have, and are motivated by objects with a median temporal extension of 2.8 years, while the Japanese might stay ahead because they appear to be guided by timeless values and motivational objects with a median temporal extension of 6.4 years. Reflecting on the impression left though the 309 interviews, the Brazilian managers must be characterized as the most hopeful group. But this research project did not include a systematic process for typifying Motivational Objects in this way.

As well as looking into the characteristics of hope and fear, in formulating a hypothesis for future research it will be important to distinguish between the processes and the effects. Time Perspective may influence the way change unfolds in a society rather than the actual outcomes: A short-term view could lead to advancement in a saw-tooth fashion through a serious of revolutions, once maintenance of the status quo becomes untenable, whereas a long-term perspective brings about more gradual change. This idea is discussed in more depth in the next section on change in complex socio-ecological systems. At this point it is important to make the distinction between scientific, technological progress and the idea that there is, and will be, long-term social progress towards civilization, which is a highly contested idea.

In a compelling essay on the history of the idea of progress Robert Nisbet (1979) mentioned four arguments that were used in the mounting criticism of the idea of progress towards the end of the last century: (1) the limits of economic advancement have been reached; (2) natural resources are being depleted; (3) there is growing disillusionment with science and technology, and (4) a waning work-ethic and growing leisure-ethic resulting in boredom. The idea of progress was more recently gutted and filleted by John Gray (2013), who has a decidedly dark view on human nature and argues that progress is a dangerous concept that results from mixing up a Socratic myth of reason and a Christian myth of salvation. According to Gray, life is not teleological and we should distrust every political movement towards an idealistic goal because they all lead to terror and bloodshed.

Gray is quite extreme in his refutal of human volition and morality, which is a stance that can sow the seeds for nihilism and apathy. On the other hand, history shows how quality of life can occasionally be improved though human interventions, for example those directed at increasing access to clean water and sanitation (Mosse and Sivan, 2003; Strang, 2004; Cameron et al., 2011), even if much more is needed than a grand plan and technological innovation alone (e.g. Sultana, 2013). As described in the interpretive framework, this research project was aligned with Archer's model of social change that distinguishes the influence of human agency. The author is inclined to recognize the value and agency of human beings, but rather than getting tangled up in a theoretical discussion about the idea of progress, be it technological, social, or moral, the focus of the next section is on the relevance of Time Perspectives for the simpler notions of transition and change.

6.3.3 Change in complex socio-ecological systems

As described in the introduction, change in the most perceptible manifestation of time itself. Archer's morphogenetic approach is based on systems theory and it shows how human interaction taking place in prior social contexts causes new contexts for subsequent interaction to emerge (Archer, 1982). Archer (1982) argues that structure and action operate over different time periods because a structure logically predates the actions that transform it and structural elaboration logically postdates those actions, as shown in Figure 6.5. Considering the nature of Chaos Theory, it may seem more self-evident than it actually is to reason that the current state of a complex system is co-determined by its past state (Cilliers, 1998). The original structural influences also extend beyond T2, so for structural elaboration to occur, that is, a transformation of states, the complex interplay between Structure and Action must reinforce the change.

Forces towards which a complex system asymptotically moves are called 'attractors' (Scheffer and Carpenter, 2003; Barnett et al., 2012). If attraction towards the current state is strong then the interplay between Structure and Action will resist a change of states. Attractors thus determine the resilience of a system. More specifically: "Resilience is the magnitude of disturbance that can be tolerated before a socio-ecological system (SES) moves to a different region of state space controlled by a different set of processes." (Carpenter et al., 2001). In the model of morphogenesis (Figure 6.5), the interplay between Structure and Action exerts temporal and directional forces by speeding-up or delaying and impelling or inhibiting the Structural Elaboration (Archer, 1982).

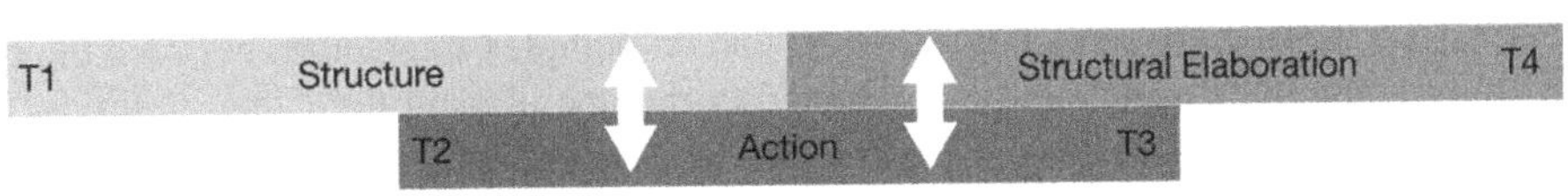

Figure 6.5 Structure and action operate over different time periods (Archer, 1982).

At this point is important to note that changes in attractors can reduce or increase the resilience of a system while having little direct influence of the actual state of that system (Scheffer and Carpenter, 2003). Such developments can, however, set-up the system for rapid change. A Tipping Point is the critical threshold beyond which a tiny perturbation can precede a massive change in conditions (Scheffer et al.,

2009). The sensitivity of a complex system to initial conditions and small disturbances is graphically captured in the cliché of a tornado caused by the flap of a butterfly's wings. The idea of a complex system shifting states under the forces of attractors and Tipping Points has been used to explain changes in aquatic systems (Scheffer et al., 1993; Baskett and Salomon, 2010) and terrestrial systems (Heffernan, 2008; Odion et al., 2009) but also to warn about planetary limits (Rockström et al., 2009) and suggest processes for restoration (Suding et al., 2004; Van Andel and Aronson, 2012).

Rather than being drawn towards alternative equilibrium conditions, socio-ecological systems tend to change following an adaptive cycle including four phases: (1) growth and exploitation; (2) conservation; (3) collapse or release, and (4) renewal or restructure (Gunderson and Holling, 2001). In this cycle the system reaches a certain state through phase 1 and remains in a dynamic balance throughout phase 2 until it collapses, perhaps via a Tipping Point, in phase 3 before restructuring in phase 4. The way social systems or organisations change obviously depends on the speed, scope, amplitude, and frequency of change in the environmental systems (Suarez and Oliva, 2005). Globalisation has been found to influence the resilience and vulnerability of socio-ecological systems through increased connectedness, speed, and spatial scope, and decreased diversity (Young et al., 2006). It would be interesting to investigate how the Motivational Space of a group, which influences the framing of problems and the nature of decisions and actions, might work as an attractor to influence the way such change unfolds. As mentioned earlier in the case of the Arab Spring, a system can be severely disrupted without a long-term view, but such a view may be needed to escape the attraction forces of the existing structure and cause enduring, directed change.

The conclusions of this research project support the idea that differences in Motivational Space across groups may generate divergent patterns in the actions that individuals from those groups take which would in turn influence the interplay between action and structure and thus generate intergroup differences in the potential for, and path to, change. It is not new to use concepts from the science of complex systems to analyse patterns and processes of change in social systems either directly or metaphorically (e.g. Dooley, 1997; Fitzgerald et al., 2002; Boyatzis, 2002). But it is easy to misuse these concepts by making loose rhetorical links (Maguire and McKelvey, 1999; Rowe and Hogarth, 2005). So rather than postulate a broader theory as an afterthought to the present study the discussion is focussed on more specific applications of a quantified Motivational Space for better defining Tipping Points and Adaptive Capacity, which are key concepts that are commonly used for Water Planning. To begin with, it is interesting to consider whether the science of Tipping Points should be focused on the matter of precisely defining that which constitutes the limit.

6.3.4 Tipping Points and Institutional Irresponsibility

Various Tipping Points have been defined with reference to Climate Change (e.g. atmospheric carbon dioxide concentration of 350 parts per million by volume), but perhaps the most current and eminent examples are the nine Planetary Limits that are meant to define a safe operating space for humanity (Rockström et al., 2009). Such limits are theoretically timeless – they are not set to occur at a fixed date but are rather indefinably dynamic deadlines that depend on complex interaction between the biophysical systems themselves and human interventions. Humanity may be playing the game of survival on a field delineated by planetary boundaries but no-one knows when the ball will cross the line, let alone when the final siren will sound. Besides this temporal fuzziness, any meaningful threshold needs to take spatio-material and social diversity into account (Folke et al., 2004; Adger, 2006). The usefulness of such theoretical limits is thus challenged by uncertainty and there are also doubts as to whether or not such predictions motivate actions at the end of the day (e.g. Dessai et al., 2009). In any case, the conclusions of this research project could be useful for clarifying the temporal dimension of Tipping Points, so the discussion of implications is limited to this aspect.

The temporal dimension of a Tipping Point; that is, when the limit will be breached, generally remains vague. Following the identification of a Tipping Point, such as the example of atmospheric carbon dioxide given above, much secondary research is directed at calculating the influence of various policy pathways (human agency) on determining the temporal proximity of when the limit will be breached (e.g. Moss et al., 2010). But this view of, and relationship with, the future is problematic for several reasons. In their scrutiny of science for the post-normal age Funtowicz and Ravetz (1993) highlighted the problems of unpredictability, incomplete control, and a plurality of legitimate perspectives. These are characteristics of Wicked Problems, as described in the Background section of this book, and they return again here to haunt those wanting to forecast Tipping Points. To address these challenges, theorists have proposed an iterative approach (Carter et al., 2007; Jones and Preston, 2011). The general idea is to act on the current knowledge and then learn from the consequences of these actions so that these insights, together with any new knowledge from other sources, can be used to direct subsequent actions. This represents a periodic investment in creating knowledge, as described in the Background (Section 1.1.5). Insight into the Foreseeable Future is useful for determining the Temporal Extent of these learning cycles, which is an implication that is described in more detail in the conclusions for setting research agendas (Boxes 6.1 & 6.2).

Learning is essential to an iterative approach, but at the same time planners need to make decisions and take actions now based on the current knowledge. The temporal proximity of Tipping Points in socio-ecological systems is determined, in part, by human actions. But the power that planners and decision makers have to cause change is far greater than the power they have to know and be mindful of the potential consequences of their decisions and actions (Adam and Groves, 2007). In other words, the players can send the ball flying right over the boundaries without realizing it until the ball is out of reach and without being reprimanded; actually, they may even collect short-term rewards for their efforts. In this way Tipping Points are commonly breached, or may inexorably be breached due to lag in the system, before the actors know what they have done and after the temporal extent of that for which they feel accountable. Water Managers struggle with this 'promethean power' on a daily basis when they invest in infrastructure and assets that have a life expectancy of 50-100 years while the median temporal extent of the Foreseeable Future is between 2 and 6 years for this group. This phenomenon is more than an unfortunate disparity; theorists who write about the politics of posterity call it 'institutional irresponsibility' and it raises serious ethical questions besides the practical ones discussed here (Beck, 1992; Day, 2013).

One of the first steps towards dealing with the mismatch between the Temporal Extent of the consequences of decisions and actions and that of what can be foreseen, is to shift the focus from 'products' to 'processes' (Adam and Groves, 2007). To operationalize this concept Adam and Groves (2007) propose the idea of a 'Timeprint'; that is, the temporal equivalent of the spatio-material Ecological Footprint (Wackernagel and Rees, 1996). The idea is to compare the temporal reach of the consequences of actions with that of (1) what can be foreseen, and (2) the established sphere of responsibility. To use an example from the water sector: The age spectrum of pumped autochthonous groundwater in the Netherlands is 10-10,000 years (Stuyfzand, 1995). Water from the sandy uplands, for instance, is at least 200 years old. The managers who use this resource have a Foreseeable Future with a median Temporal Extension of 2 years. A comparison of Timeprints suggests that institutional irresponsibility may be at play. The next question is whether or not those who use this resource are taking actions to ensure that it will be recharged for at least 200 years into the future.

Besides groundwater, boreholes can cause contamination of soil or surface water with brine and other compounds from deeper strata. This has raised concerns in connection with geothermal heat pumps and, more recently, shale gas exploitation. Drillers assure us that the boreholes are sealed after use, but for how long? Can managers whose decisions and actions are motivated by objects with a median Temporal Extension of 2 years convincingly guarantee that boreholes will be sealed indefinitely? As described in the Introduction, in answering this question it is important not to lump Motivational Objects, which influence decisions and actions, with ineffectual wishful thinking. The question is not only if the actors feel responsible but also whether or not they act responsibly. More specific research would be needed to determine whether or not this is actually the case, but the Timeprints give some grounds for doubt. And if this question seems weighty, imagine what the concept of a Timeprint means for those claiming the right to dump radioactive waste underground (Streffer et al., 2012).

In moving the focus of resource management beyond certainty and control water planners need to recognize that "uncertainty of potential outcomes cannot absolve producers of long-term, open-ended impacts from responsibility to those affected in remote futures and places." (Adam and Groves, 2007, p.185). But what tools are water planners left with if scientific prediction and economic risk calculation alone are insufficient? The theoretical answer is to link action directly to responsibility and ethics. Post-normal science may take uncertainty and ambiguity into account explicitly (Funtowicz and Ravetz, 1993), but integration of political values into scientific investigations is problematic. The risks associated with routing the link between action and ethics via the science system is that scientists misuse their authority (e.g. Jasanoff, 1994) and that this authority becomes undermined (Pielke, 2007; Hulme, 2009). Leadership and morality cannot be substituted by certainty and foreseen necessity. The burden of wicked problems sees managers and policy makers passing the hot potato of responsibility for decisions actions to scientists. But accepting this task is foolish since the science system cannot determine what is right or wrong or generate knowledge about the unknowable. Water planners need to realize that their decisions may require forgiveness from successors and act in a social context of indebtedness (Adam and Groves, 2007). From a practical perspective, the idea of a Timeprint is explored in the next section.

6.3.5 Windows of opportunity

A Timeprint represents the temporal reach of actions and raises the question of whether the consequences of an action extend beyond the temporal extent of the responsibility taken for that action (Adam and Groves, 2007). The conclusions characterize the Foreseeable Future of water professionals worldwide and quantify it in terms of Temporal Extent and Certainty. This section is dedicated to evaluating the practical implications of these time horizons for water planning. To discuss a system's dynamics the system boundaries first need to be defined. For analysing processes of change for actors in the water sector it is useful to define the system boundaries in terms of the sphere of influence (Levine, 1972; Gharajedaghi, 1999; Wood, 2011).

That which is located within the spatial and conceptual realm over which an actor has significant cultural, economic, political, or physical control is referred to as an Internal System. The External Systems thus include the rest of the world over which the actor has no influence. There is grey area that makes the boundary between the Internal and External fuzzy, which is referred to as the Transactional Space. An actor does not have direct control over the Transactional Space but may, for instance through lobbying, influence other people or physical systems to change in a particular way. The Internal Systems are conceived as surrounded by External Systems with the Transactional Space in between (Figure 6.6).

The systems in Figure 6.6 are not static but dynamic: both concerning the location of the system boundaries and the interaction between the Internal and the External Systems. Planning refers to making decisions to take actions directed at changing or maintaining certain properties of the Internal Systems and the Transactional Space. This directed change may be proactive by anticipating changes in the External Systems and Transactional Space, changes that pose threats or create opportunities for the Internal Systems, or reactive, by responding to the changes once they have occurred. Some planning strategies require flexibility but as the costs of adaptation increase, as with investments in infrastructure, so too do the benefits of preparedness.

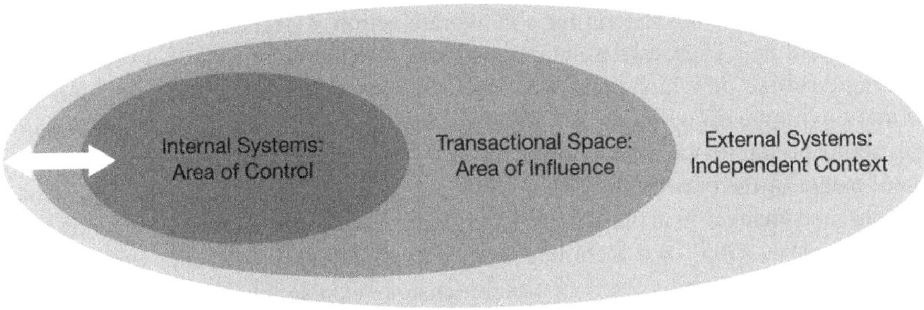

Figure 6.6 Spheres of control and influence (Gharajedaghi, 1999).

As described in the Introduction, water organizations are currently struggling with the transition to a more adaptive approach to planning. The consequences of a particular change event for the Internal Systems can be seen positively, in terms of capacity, or negatively, in terms of vulnerability (Field et al., 2012). Adaptive Capacity is generally understood in terms of the Internal System's ability to learn and adjust to changes (Folke et al., 2010). Adaptive planning involves some 'humility' regarding the future. It echoes the idea of 'indebtedness to successors' proposed by Adam and Groves (2007). Proactively scanning the horizon of the Foreseeable Future for changes in External Systems and the Transactional Space is enhanced and extended by 'systematic learning' (Pahl-Wostl et al., 2011), one of the most central processes in an adaptive approach. Knowing the temporal extent of the Foreseeable Future perceived by water professionals has some very practical consequences for adaptive capacity.

Adaptive capacity can be divided into three types according to whether the capabilities are useful before, during, or after the event that causes change (Adger, 2006). Pre-impact adaptive capacity refers to the ability to learn, anticipate, and prepare. This type of adaptive capacity is often translated into terms of mitigation and planning. After change has been triggered, during the initial and most rapid course of change, adaptive capacity is represented by the potential to immediately react by taking advantage of opportunities that emerge and preventing or absorbing adverse effects. Post-impact adaptive capacity involves the ability to recover and evolve to reduce future vulnerability. Insight into the Foreseeable Future of water professionals is particularly relevant for pre-impact adaptive capacity and thus planning.

Planning is basically defined as "the design of a desired future and of effective ways of bringing it about" (Ackoff, 1970). Three elements comprise this definition: (1) 'design', which implies purposeful thought; (2) 'a future', which is a Motivational Object, and (3) 'ways of bringing it about', which are actions directed at changing or maintaining certain properties of the Internal Systems and the Transactional Space (Figure 6.6). Geels and Schot (2007) argue that no transition is planned and coordinated from the outset and that the convergence of the plans and activities that emerges during the process itself is most essential. Insight into the ambiguity in the Motivational Space of these various actors is likely to be advantageous in facilitating this convergence. And there are likely to be various ways of applying these insights. One striking possibility is discussed here.

To explain how a quantified Foreseeable Future can be used for planning a static model is described, which is then made increasingly dynamic. Consider that the temporal dimension of a change in the External System is used to set the time-limit (i) for the use of pre-impact adaptive capacity. After a certain change has taken place the actions directed at dealing with that change are no longer referred to as planning. Next it is assumed that a certain amount of time is needed to take action towards adapting; that is, the 'ways of bringing it about'. To meet the criteria for a deliberate action, a decision must therefore be made by a certain 'planning deadline'. These two assumptions leave a 'window of time' between the present 'now' and the 'planning deadline' in which purposeful thought can be directed at gathering and using decision support information (Figure 6.7). Pre-impact adaptive capacity may thus be divided into a 'thinking period' (n - p) and an 'acting period' (p - i). These horizons represent an analytical tool for exposing the importance of the temporal extent of the Foreseeable Future and are not meant to represent the reality of interrelated decisions, actions, and changes in actual Internal and External systems. Real-life decisions are also seldom made rationally (Ainslie, 2005). But including these complexities in the model would distract from the point to be illustrated, which is why the temporal dimension is isolated here.

Supposing the four simple time horizons shown in Figure 6.7 exist, several propositions are seen to govern the dynamics:

1. The temporal extent of each horizon changes over time and is unknown.
2. Each period between horizons may be reduced to zero, in which case all four horizons coincide at the present now (n).
3. All actions are irreversible, so the Planning Deadline (p) represents the limit for gathering decision support input for a given action: the period (n – p) logically predates (p – i).
4. There is no point in planning to act if there is no time to act: the period (n – p) equals zero if (p – i) is zero.
5. The Temporal Extent of the Foreseeable Future is = 0 and may be less than the periods (n – p); (p – i), or even (n – i).

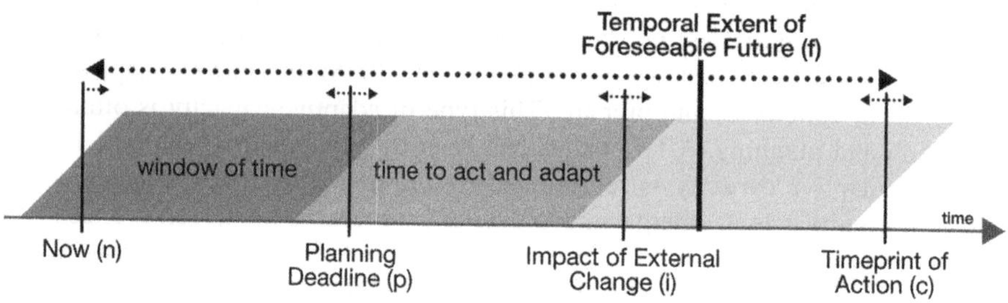

Figure 6.7 The dynamics of time horizons for water planning.

The amount of time between each of these limits is thus ultimately determined by the horizon of the Foreseeable Future. If the Foreseeable Future (f) has a Temporal Extent that is greater than or equal to horizon (i) then the actor has time to act and adapt plus a window of time for planning. Any shortening of horizon (f) results in reduction of the period (n – p) and/or (p – i). As horizon (f) approaches (n) the actor becomes increasingly reactive and may have little or no time to adapt to change. The perceived extent of this horizon was found to vary significantly across National Cultures and even more so across Professional Roles. Time Perspectives can be purposefully altered (Zimbardo and Boyd, 2008), but the predictability of the External Systems is beyond the sphere of influence (Figure 6.6). Change in External Systems can happen in quite different ways, which in turn demand different capacities of the actors who are impacted. Suarez and Oliva (2005) defined five types of change (Table 6.3) that are characterized by four attributes:

Frequency: Number of disturbances per unit time.
Amplitude: Magnitude of the deviation from the initial conditions.

Speed: The rate at which this deviation occurs.
Scope: The number of dimensions of the systems that are affected.

The extent of a future that can be accurately foreseen is likely to be shorter in the context of, for example, disruptive change than if regular change is occurring. But what is practically more critical, and also at the core of this study, is how foreseeable the actor perceives the future to be. In the context of disruptive change, a more proficient actor who has more strategic intelligence information could (accurately) perceive the future to be less foreseeable than an ignorant and incompetent actor. This type of change creeps up slowly, before reaching a critical mass or tipping point, and can thus easily go unnoticed until it is too late. Until now the relative proficiency of actors was excluded from the analysis because the idea of assessing the realism of an individual's Time Perspective is inconsistent with the notion that the future is unknowable. But to translate the conclusions into practical applications it is necessary to address factors that can be influenced. And that which can be influenced can also be performed more or less proficiently.

The proficiency of an actor, in water planning or setting research agenda's, may be assessed from various perspectives. Regarding actor proficiencies in addressing the Foreseeable Future; if all other aspects are considered equal, an actor who motivates their decisions and actions based on objects having greater Temporal Extent, has advantages over an actor who is focussing on the short-term alone. This has been proven on the individual level when it comes to education, the use of drugs, and prosperity later in life (Zimbardo and Boyd, 2008). Levels of affluence and economic growth have also been found to correlate with long-term orientation on the national scale (Hofstede et al., 2010). Now looking again at Figure 6.7, the horizon (f) can be set at 3.5 years to represent the median Temporal Extent of Managers in the water sector worldwide. The Timeprints of the actions of this group (c) have been shown to extend well beyond this limit considering, for example, the service life of the assets that are managed. But the focus here is on the consequences of the Temporal Extent of the Foreseeable Future for the Internal Systems and not External Systems, such as Earth's atmosphere and the climate.

Table 6.3 Types of change in the External Systems (adapted from Suarez and Oliva, 2005).

Frequency	Amplitude	Speed	Scope	Type of change
Low	Low	Low	Low	Regular
High	Low	High	Low	Hyper-turbulence
Low	High	High	Low	Specific Shock
Low	High	Low	Low	Disruptive
Low	High	High	High	Avalanche

While regular change is occurring, extrapolation of past change might be used successfully to predict future change and plan accordingly. The projections of domestic water demand in the Netherlands from 1950 until 1990 are a good example of this situation – they worked well during this period. This approach could also work roughly during periods of hyper-turbulence when there is frequent and fast change but the deviation from the initial conditions is low. Under both of these conditions the 'thinking period' (n -p) can be negligible and the actor can focus on actions. But the 'thinking period' (n - p) and an 'acting period' (p - i), which for Managers generally sums to a maximum of 3.5 years, could be catastrophic when it comes to specific shocks, avalanches, and disruptive change.

It is important to note that the Foreseeable Future is defined by the Temporal Extent of the objects that motivate and actor's decisions and actions. A Manager may recognize and talk about temporally distant objects, but if these objects are not used to motivate decisions and actions then they are ineffectual. One reason for this could be uncertainty. This is what makes contingency planning a useful approach. The role recommended for Scientists in the summary of implications and applications is to signal emerging opportunities and threats and to make contingency plans. Because if the Managers begin planning after an object reaches the Temporal Extent of their Foreseeable Future then their adaptive capacity is generally limited. This limitation can be addressed, to some extent, by taking multiple scenarios and contingencies into account throughout the hierarchy of planning processes (e.g. Table 6.2).

Assuming equal proficiency of actors in their accuracy predicting the future, and that no contingency plans have been made for the impact in question (i), adaptive capacity is logically related to the Temporal Extent of the Foreseeable Future (f). To use the example of Managers: A Dutch Manager with a median (f) of 2 years, would have time for thinking (n - p) and acting (p - i), prior to the impact occurring, which is half that of a Brazilian Manager with a median (f) of 4 years. The 'window of time' for planning, testing response strategies and adapting, is thus limited by the Temporal Extent of the Foreseeable Future. One way of maximizing this 'window of time' is to minimize the need to adapt and/ or the time needed for acting and adapting. To do this the actor can invest in robustness and flexibility (Meijer, 2007). From a practical perspective these qualities can be achieved by, for example, building redundancies and multiple barriers into the Internal System, using modules rather than a single component system, and minimizing the long-term investments by using components that can be cheaply replaced.

There is, however, a limit as to what can be done to minimize the need to adapt and/or the time needed for adapting. And there are trade-offs between the various strategies (Meijer, 2007). These options are well known and already widely applied in the water sector. The purpose of focussing on the temporal dimension of these activities is to underline the importance of the Temporal Extent of the Foreseeable Future in determining the adaptive capacity. This conclusion is, for example, relevant for the United Nations frameworks for action, such as the Millennium Development Goals and the World Water Assessment Programme. From a temporal perspective Type 1 and 2 countries have less adaptive capacity than Type 3 and 4 countries (Figure 6.1), and may thus be more vulnerable to disruptive change. All the same, other factors like the ability to recover from a harmful event are at least as important.

Further to anticipating the change in the External Systems and the consequences of the change for the Internal Systems, is being able to apply an iterative, adaptive approach to planning. This means using the 'window of time' for systematic learning from previous actions. To be more realistic Figure 6.7 would need to include multiple interrelated potential impacts and associated windows of time. One way of tracking these various horizons is to add the dimension of 'velocity' to limits (e.g. Loarie et al., 2009). The secret to avoiding collapse of the internal system in the long-term may be to continue the active horizon scanning, scenario planning, and design of contingency plans even during periods of regular, predictable change. This is likely to maximise the adaptive capacity for when specific shocks, avalanches, and disruptive changes occur.

The discussion to this point has focussed on how changes in the External Systems can impact Internal Systems. But actors are not helpless victims of fate; they have control over the Internal Systems and influence on the Transactional Space. In addition to taking action towards conserving the current state of these systems, actors can proactively design their futures in part. A vision represents the desired future state of the Internal Systems and, to some extent, the Transactional Space. Actors in the water sector may, for example, aim for a '20% increase in the population served with drinking water by 2020' or '10% reduction in leakage losses in the distribution network by 2015' or 'water treatment without chlorine by 2030'. As with the moon landing, none of these visions is necessary, per se. One of the main uses of a vision is to be a source of inspiration, and this association invokes the idea of hope. For planners and agenda setters, the foremost recommendation is to use the conclusions to project visions on horizons that motivate people and thus proactively affect change. Further recommendations based on the conclusions are summarized below.

6.4 LIST OF RECOMMENDATIONS

- Invest more time and resources in the framing of problems. This is particularly important for participatory planning and transdisciplinary research because these activities involve people from different professions who tend to have quite different Time Perspectives. The results of this research project indicate that part of the ambiguity in problem perception is likely to originate from differences in the Temporal Extent of the future that is considered foreseeable by different (groups of) people. Scientists tend to be motivated by problems and goals on more distant time horizons than managers and practical workers. People who have different professional roles are also likely to have different 'desired futures' that a planning process could be directed at 'bringing about'. The process of collectively framing and reframing problems and goals can result in learning and change that is an essential part of the response. The recommendations for setting research agendas are similar to those for planning in this regard: The framing of research problems needs to become a greater part of research itself.
- Collectively and explicitly map the Motivational Space from the outset of all planning projects, especially in terms of Temporal Extent. The Foreseeable Future Multi-measure Method was designed for this purpose. Characterizing and considering diversity in Motivational Space is particularly important for participatory planning, which involves actors who have different professional roles. Once Motivational Space has been mapped it can be used to make plans that are more likely to influence the decisions and actions of the actors. A planning process can be adapted to address the diversity in Motivational Space by, for example, including milestones or objectives that have temporal targets which are tailored to the Time Perspectives of the relevant actors. Alternatively, the membership of the planning team can be specifically chosen to make plans for a certain time horizon.
- Collectively and explicitly map the Motivational Space of actors involved in setting research agendas. A research agenda can be adapted to accommodate the diversity in Motivational Space of the people in a transdisciplinary team or the team of agenda setters can be chosen to programme research of a certain type. Both alternatives explicitly attend to the Motivational Space of those involved. The former option might be favoured over the latter in Europe at present due to democratization of scientific knowledge and the demand for practical innovations. Research funding is increasingly directed at developing market-ready technological tools in collaborative projects between universities, government, and industry partners. But rather than compromising all research to accommodate the Motivational Space of all professional roles, which is likely to limit the problems to those with a time horizon of up to 4 years, it is recommended that a portion of the research funding also be allocated to research on different temporal scales.
- Compare the time perspectives of actors with the time quality of their actions as an indicator of sustainability. More specifically, the Motivational Space of a group of actors can be compared with the sphere of influence, the sphere of established responsibility, and the Timeprints of (planned) activities. If the Timeprints of activities exceed the Temporal Extent of the sphere of established responsibility then those activities are ethically questionable. Similarly, if the Temporal Extent of the sphere of influence is greater than that of the Motivational Space then the actors are making decisions and taking actions based on a temporally filtered fraction of the potential outcomes of those activities. The future is unknown, so there is prospective uncertainty about the Timeprints of some activities. But other Timeprints, such as the half-life of nuclear waste and the recharge rate

of an aquifer, are relatively predictable. It is likely to be most meaningful and effective to compare these dimensions in cooperation with the relevant actors, including explicit discussions about the uncertainties, rather than trying to calculate them or judge from a distance. This recommendation can be generalized as follows: avoid using age-old and irreplaceable resources for short-term gains.

- Define the Temporal Extent of the Motivational Space of decision makers and scan beyond that time horizon for emerging opportunities and threats to maximize adaptive capacity. Systematic horizon scanning can provide insights that inspire decision makers to look beyond day- to-day issues and short-term politics. A transdisciplinary strategic learning alliance can be used to facilitate co-learning across professional roles and expand the scope of choices available to decision makers. But more work is needed to find ways of accommodating diversity in the level of knowledge and the learning capacities of members. To feed a strategic learning alliance with new insights, scientists need to be employed to scan horizons with a Temporal Extent of more than 5 years into the future and to develop signposts to signal potential tipping points. If all other aspects are considered equal, actors with the ability to motivate their decisions and actions using more temporally distant objects have more capacity to adapt to change than those who are focussed on the short-term alone.
- To influence the decisions and actions of managers in the water sector scientists need to scale insights about long term change down to Motivational Objects with a Temporal Extent of 2-5 years. Downscaling is, however, notoriously problematic. Patterns discovered by means of analysis on a long-term scale may differ from those that emerge on the short-term. Similarly, interventions taken in response to short-term problems can generate new long-term problems. The hierarchies and scales apparent in the biophysical systems being managed might seem to be most relevant to the problems at hand but, when it comes to influencing behaviour, the human scale of Motivational Space should not be overlooked. Invest in social and psychological research to improve the human side of scaling wicked problems, like Climate Change, down to the Motivational Space of decision makers.
- Incorporate the Temporal Extent of the Foreseeable Future into asset management methods and tools. To monitor and maintain existing infrastructure and make intelligent new investments an understanding of the Motivational Space of decision makers is useful. The functional service life of a system or component can only be assumed to extend as far as the Foreseeable Future. In the case of the water sector the economic and technical service life of most infrastructural assets is longer than the Temporal Extent of the Foreseeable Future. One response could be to investigate possibilities for investing in infrastructure that has a shorter technical service life. This is likely to only be economical for above ground infrastructure for which the installation costs are relatively low. A more adaptive response might be to ensure that the components are demountable, re-usable, and/or recyclable. This idea also complements new ideals for material flows according to the principles of cradle-to-cradle design and a circular economy.
- Design visions on time horizons that motivate people to inspire positive change. A long term horizon can be intentionally used to distance people from their everyday routines and enable them to openly explore various scenarios by neutralizing the passion with which they defend perceived stakes. The other side of the coin, as empirically evidenced by this research project, is that most of the objects that influence decisions and actions are perceived as having a 70 percent likelihood of occurring within the next 4 years. It seems rational to assume that a vision with these characteristics is also most likely to affect change.

References

Abrams D., Wetherell M., Cochrane S., Hogg M. A. and Turner J. C. (1990). Knowing what to think by knowing who you are: self-categorization and the nature of norm formation, conformity and group polarization. *British Journal of Social Psychology,* **29**(2), 97-119.

Ackoff R. L. (1970). A Concept of Corporate Planning. Wiley, New York, USA.

Adam B. (2004). Time. Polity Press, Cambridge, UK.

Adam B. and Groves C. (2007). Future Matters: Action, Knowledge, Ethics. Brill, Leiden, The Netherlands.

Adams J. and Nettle D. (2009). Time perspective, personality and smoking, body mass, and physical activity: an empirical study. *British Journal of Health Psychology,* **14,** 83-105.

Adger W. N. (2006). Vulnerability. *Global Environmental Change,* **16**(2), 268-281.

Agrawal A. and Lemos M. (2007). A greener revolution in making? Environmental governance in the 21st century. *Environment,* **49**(5), 36-45.

Ainslie G. (2005). Précis of breakdown of will. *Behavioral and Brain Sciences,* **28,** 635-673.

Al-Amoudi I. and Willmott H. (2011). Where constructionism and critical realism converge: interrogating the domain of epistemological relativism. *Organization Studies,* **32**(1), 27-46.

Allan J. A. (1998). Virtual water: a strategic resource global solutions to regional deficits. *Groundwater,* **36**(4), 545-546.

Archer M. S. (1982). Morphogenesis versus structuration: on combining structure and action. *The British Journal of Sociology,* **33**(4), 55-483.

Arnold G., Bos H., Doef R., Goud R., Kielen N. and Van Luijn F. (2011). Water Management in the Netherlands. Ministry of Infrastructure and Environment, Directorate-General Water and Rijkswaterstaat, Centre for Water Management.

Azariadis C. (1981). Self-fulfilling prophecies. *Journal of Economic Theory,* **25**, 380-396.

Baggelaar P. K. and Geudens P. J. J. G. (2011). Prognosis of National Domestic Water Demand 2025. Association of Dutch Water Companies (Vewin), Rijswijk, The Netherlands.

Barber B. (1984). Strong Democracy: Participatory Politics for a New Age. University of California Press, Berkeley, CA, USA.

Barnett W. A., Serletis A. and Serletis D. (2012). Nonlinear and complex dynamics in economics. Forthcoming in *Macroeconomic Dynamics,* September 11, 2012.

Barney D. (2004). The Network Society. Polity Press, Cambridge, UK.

Baskett M. L. and Salomon A. K. (2010). Recruitment facilitation can drive alternative states on temperate reefs. *Ecology*, **91**, 1763-1773.

Beck U. (1992). Risk Society: Towards a New Modernity. Sage Publications.

Bellamy R. and Hulme M. (2011). Beyond the tipping point: understanding perceptions of abrupt climate change and their implications. *Weather, Climate, and Society,* **3**(1), 48-60.

Benedict R. (1989). The Chrysanthemum and the Sword: Patterns of Japanese Culture. Houghton Mifflin Harcourt, Boston, MA, USA.

Benjamin A. C. (1981). Ideas of time in the history of philosophy. In: The Voices of Time: A Cooperative Survey of Man's Views of Time as Expressed by the Sciences and by the Humanities, J. T. Fraser (ed.), The University of Massachusetts Press, Amherst, MA, USA.

Berkhout F., Smith A. and Stirling A. (2004). Socio-technological regimes and transition contexts. In: System Innovation and the Transition to Sustainability: Theory, Evidence and Policy, B. Elzen, F. W. Geels and K. Green (eds.), Edward Elgar, Cheltenham, UK, pp. 48-75.

Beuken R. H. S. and Mesman G. A. M. (2011). Technical Service Life for Groups of Pipelines: Towards a Basis for Investment Policy, KWR Report number BTO 2011:038, The Netherlands.

Bhaskar R. (1978). A Realist Theory of Science. Harvester Press, Hassocks, UK.

Bian Y. (1994). Work and Inequality in Urban China. State University of New York Press, Albany, NY, USA.

Bickel W. K., Jarmolowicz D. P., Mueller E. T., Koffarnus M. N. and Gatchalian K. M. (2012). Excessive discounting of delayed reinforcers as a trans-disease process contributing to addiction and other disease-related vulnerabilities: emerging evidence. *Pharmacology and Therapeutics,* **134**(3), 287-297.

Bickel W. K., Pitcock J. A., Yi R. and Angtuaco E. J. C. (2009). Congruence of BOLD response across intertemporal choice conditions: fictive and real money gains and losses. *The Journal of Neuroscience,* **29**(27), 8839–8846.

Biswas A. K. (2004). Integrated water resources management: a reassessment. *Water International,* **29**(2), 248-256.

Block R. A. (1990). Cognitive Models of Psychological Time. Lawrence Erlbaum Associates, Hillsdale, MI, USA.

Boniecki G. (1980). What are the limits to man's time and space perspectives? Toward a definition of a realistic planning horizon. *Technological Forecasting and Social Change,* **17**, 161-175.

Bontempo R. (1993). Translation fidelity of psychological scales: an item response theory analysis of an individualism-collectivism scale. *Journal of Cross-Cultural Psychology,* **24**, 149-166.

Boroditsky L. (2011). How languages construct time. In: Space, Time and Number in the Brain: Searching for the Foundations of Mathematical Thought, S. Dehaene and E. M. Brannon (eds.), Academic Press, London, UK, pp. 333-341.

Bouwen R. and Taillieu T. (2004). Multi-party collaboration as social learning for interdependence: developing relational knowing for sustainable natural resource management. *Journal of Community and Applied Social Psychology,* **14**(3), 137-153.

Boyatzis R. E. (2006). An overview of intentional change from a complexity perspective. *Journal of Management Development,* **25**(7), 607-623.

Brody S. D. and Highfield W. E. (2005). Does planning work?: testing the implementation of local environmental planning in Florida. *Journal of the American Planning Association,* **71**(2), 159-175.

Brown R. B. and Herring R. (1998). The circles of time: an exploratory study in measuring temporal perceptions within organizations. *Journal of Managerial Psychology,* **13**(8), 580-602.

Brugnach M., Dewulf A., Pahl-Wostl C. and Taillieu T. (2012). Toward a relational concept of uncertainty: about knowing too little, knowing too differently, and accepting not to know. *Ecology and Society,* **13**(2), 30.

Brugnach M. and Ingram H. (2012). Ambiguity: the challenge of knowing and deciding together. *Environmental Science and Policy,* **15**(1), 60-71.

Bruno J. E. (1995). Doing time—Killing time at school: An examination of the perceptions and allocations of time among teacher-defined at-risk students. *The Urban Review* **27**(2): 101-120.

Bruns B. (2003). Water Tenure Reform: Developing an Extended Ladder of Participation, Paper presented at Politics of the Commons: Articulating Development and Strengthening Local Practices, Chiang Mai, Thailand, July 11-14.

Buchan N. R., Grimalda G., Wilson R., Brewer M., Fatas E. and Foddy M. (2008). Globalization and human cooperation. *Proceedings of the National Academy of Sciences USA,* **106**, 4138-4142.

Buizer M., Arts B. and Kok K. (2011). Governance, scale and the environment: the importance of recognizing knowledge claims in transdisciplinary arenas. *Ecology and Society,* **16**(1). http://researchrepository.murdoch.edu.au/6407/1/governance_scale_and_the_environment.pdf

Burby R. J. (2003). Making plans that matter: citizen involvement and government action. *Journal of the American Planning Association,* **69**(1), 33-49.

Burke P. (2005). History and Social Theory. Polity Press, Cambridge, UK, pp. 120-140.

Busby J. S., Alcock R. E. and MacGillivray B. H. (2009). Interrupting the social amplification of risk process: a case study in collective emissions reduction. *Environmental Science and Policy,* **12**, 297-308.

Cameron J., Jagals P., Hunter P. R., Pedley S. and Pond K. (2011). Economic assessments of small-scale drinking-water interventions in pursuit of MDG target 7C. *Science of the Total Environment,* **410**, 8-15.

Campbell H. and Marshall R. (1999). Ethical frameworks and planning theory. *International Journal of Urban and Regional Research,* **23**(3), 464-478.

Carolan M. S. (2005). Society, biology, and ecology: bringing nature back into sociology's disciplinary narrative through critical realism. *Organization Environment,* **18**, 393-421.

Carpenter S., Walker B., Anderies J. M. and Abel N. (2001). From metaphor to measurement: resilience of what to what? *Ecosystems,* **4**(8), 765-781.

Carter T. R., Jones R. N., Lu X., Bhadwal S., Conde C., Mearns L. O., O'Neill B. C., Rounsevell M. D. A. and Zurek M. B. (2007). New assessment methods and the characterisation of future conditions. In: Climate Change 2007: Impacts, Adaptation and Vulnerability. Contribution of Working Group II to the Fourth Assessment Report of the Intergovernmental Panel on Climate Change, M. L. Parry, O. F. Canziani, J. P. Palutikof, P. J. Van der Linden and C. E. Hanson (eds.), Cambridge University Press, Cambridge, UK, pp. 133-171.

Casasanto D. and Boroditsky L. (2008). Time in the mind: using space to think about time. *Cognition,* **106**, 579-593.

Castells M. (1996). The Rise of the Network Society. The Information Age: Economy, Society and Culture, Vol. 1.Blackwell Publishers, Malden, MA, USA.

Charlton S. R., Gossett B. D. and Charlton V. A. (2011). Beyond the shadow of a trait: understanding discounting through item-level analysis of personality scales. *The Psychological Record,* **61**, 583-598.

Cheng A. S. and Daniels S. E. (2005). Getting to "we": examining the relationship between geographic scale and ingroup emergence in collaborative watershed planning. *Human Ecology Review,* **12**(1), 30-43.

Christoforou T. (2003). The precautionary principle in European community law and science. In: Precaution, Environmental Science, and Preventive Public Policy, J. A. Tickner (ed.), Island Press, London, UK, pp. 241-262.

Cilliers P. (1998). Complexity and Postmodernism: Understanding Complex Systems. Routledge, London, UK.

Cinyabuguma M., Page T. and Putterman L. (2005). Cooperation under the threat of expulsion in a public goods experiment. *Journal of Public Economics,* **89**, 1421-1435.

Cohen J. (1988). Statistical Power Analysis for the Behavioral Sciences, 2nd edn, Lawrence Earlbaum Associates, Hillsdale, MI, USA.

Cohen J., Hansel C. E. M. and Sylvester J. (1954). An experimental study of comparative judgments of time. *British Journal of Psychology,* **55**, 108-114.

Cottle T. J. (1967). The circles test: an investigation of perceptions of temporal relatedness and dominance. *Journal of Projective Techniques and Personality Assessment,* **31**(5), 58-71.

Cottle T. J. and Pleck J. H. (1969). Linear estimations of temporal extensions: the effect of age, sex, and social class. *Journal of Protective Techniques and Personality Assessment,* **33**, 81-93.

Crotty M. (1998). The Foundations of Social Research: Meaning and Perspective in the Research Process. Allen and Unwin, St Leonards, NSW.

Csikszentmihalyi M. (1975). Beyond Boredom and Anxiety. Jossey-Bass, San Francisco, CA, USA.

Currie-Alder B. (2007). Unpacking participatory NRM: distinguishing resource capture from democratic governance. In: Multi Stakeholder Platforms for Integrated Water Management, J. Warner (ed.), Ashgate, Burlington, VT, USA, p. 245.

D'Alessio M., Guarino A., De Pascalis V. and Zimbardo P. G. (2003). Testing Zimbardo's Stanford Time Perspective Inventory (STPI) – short form. An Italian study. *Time and Society,* **12**(2), 333–347.

Davoudi S. (2001). Planning and the twin discourses of sustainability. In: Planning for a Sustainable Future, A. Layard, S. Davoudi and S. Batty (eds.), Spon, London, UK, p. 81.

Day N. (2013). The politics of posterity: expertise and long-range decision-making. In: Future Directions for Scientific Advice in Whitehall, R. Doubleday and J. Wilsdon (eds.), Sciencewise-ERC, Didcot, UK, pp. 106-114.

De Graaf R. and Der Brugge R. V. (2010). Transforming water infrastructure by linking water management and urban renewal in Rotterdam. *Technological Forecasting and Social Change*, **77**(8), 1282-1291.

Dessai S., Hulme M., Lempert R. and Pielke Jr R. (2009). Climate prediction: a limit to adaptation? In: Adapting to Climate Change: Thresholds, Values, Governance, W. N. Adger, I. Lorenzoni, and K. L. O'Brien (eds.), Cambridge University Press, Cambridge, UK.

Dessai S., Lu X. and Risbey J. S. (2005). On the role of climate scenarios for adaptation planning. *Global Environmental Change,* **15**, 87-97.

De Volder M. L. and Lens W. (1982). Academic achievement and future time perspective as a cognitive-motivational concept. *Journal of Personality and Social Psychology*, **42**(3), 566-571.

Dewulf A., Craps M., Bouwen R., Taillieu T. and Pahl-Wostl, C. (2005). Integrated management of natural resources: dealing with ambiguous issues, multiple actors and diverging frames. *Water Science and Technology,* **52**(6), 115-124.

Diamond J. (2005). Collapse: *How societies choose to fail or succeed.* Penguin.

Doob L. W. (1971). Patterning of Time. Yale University Press, New Haven, CI, USA.

Dooley K. J. (1997). A complex adaptive systems model of organization change. *Nonlinear Dynamics, Psychology, and Life Sciences,* **1**(1), 69-97.

D'Oro G. (ed.) (2013). Reasons and Causes: Causalism and Non-Causalism in the Philosophy of Action. Palgrave Macmillan, Basingstoke, UK.

Dreber A. and Nowak M. A. (2008). Gambling for global goods. *Proceedings of the National Academy of Sciences USA,* **105**, 2261-2262.

Eberhard R., Robinson C. J., Waterhouse J., Parslow J., Hart B., Grayson R. and Taylor B. (2009). Adaptive management for water quality planning – from theory to practice. *Marine and Freshwater Research,* **60**(11), 1189-1195.

Ellingson J. E., Sackett P. R. and Hough L. M. (1999). Social desirability corrections in personality measurement: issues of applicant comparison and construct validity. *Journal of Applied Psychology,* **84**(2), 155-166.

Elliott M., Gray B. and Lewicki R. (2003). Lessons learned about the framing of intractable environmental conflicts. In: Making Sense of Intractable Environmental Conflicts: Concepts and Cases, R. Lewicki, B. Gray and M. Elliott (eds.), Island Press, Washington, D.C., USA, pp. 409-436.

Ellsberg D. (1961). Risk, ambiguity, and the savage axioms. *Quarterly Journal of Economics,* **75**(4), 643-669.

Enserink B., Patel M., Kranz N. and Maestu J. (2007). Cultural factors as co-determinants of participation in river basin management. *Ecology and Society,* **12**(2), 24.

Etzkowitz H. (2012). Triple helix clusters: boundary permeability at university–industry–government interfaces as a regional innovation strategy. *Environment and Planning C: Government and Policy,* **30**, 766-779.

Etzkowitz H., De Mello C. and Manoel J. (2004). The rise of a triple helix culture: innovation in Brazilian economic and social development. *International Journal of Technology Management and Sustainable Development,* **2**(3), 159-171.

Field C. B., Barros V., Stocker T. F. and Dahe Q. (eds.) (2012). Managing the Risks of Extreme Events and Disasters to Advance Climate Change Adaptation, Special Report of the Intergovernmental Panel on Climate Change, Cambridge University Press, Cambridge, UK.

Fitzgerald L. A. and Van Eijnatten F. M. (2002). Chaos speak: a glossary of chaordic terms and phrases. *Journal of Organizational Change Management,* **15**(4), 412-423.

Folke C., Carpenter S., Elmqvist T., Gunderson L., Holling C. S. and Walker B. (2002). Resilience and sustainable development: building adaptive capacity in a world of transformations. *AMBIO: A Journal of the Human Environment,* **31**(5), 437-440.

Folke C., Carpenter S. R., Walker B., Scheffer M., Chapin T. and Rockström J. (2010). Resilience thinking: integrating resilience, adaptability and transformability. *Ecology and Society,* **15**(4), 20.

Folke C., Carpenter S. R., Walker B., Scheffer M., Elmqvist T., Gunderson L. and Holling C. S. (2004). Regime shifts, resilience, and biodiversity in ecosystem management. *Annual Review of Ecology Evolution and Systematics,* **35**, 557-581.

Frederick S., Loewenstein G. and O'donoghue T. (2002). Time discounting and time preference: a critical review. *Journal of Economic Literature,* **40**(2), 351-401.

Frijns J., Büscher C., Segrave A. and Van der Zouwen M. (2013). Dealing with future challenges: a social learning alliance in the Dutch water sector. *Water Policy,* **15**(2), 212-222.

Funtowicz S. O. and Ravetz J. R. (1993). Science for the post-normal age. *Futures*, **25**(7), 739-755.

Geels F. W. and Schot J. (2007). Typology of sociotechnical transition pathways. *Research Policy,* **36**(3), 399-417.

Geertz C. (1973). The Interpretation of Cultures. Basic Books, New York, NY, USA.

Gell A. (1996). The Anthropology of Time. Cultural Constructions of Temporal Maps and Images. Berg, Oxford, UK.

Genus A. and Coles A. M. (2008). Rethinking the multi-level perspective of technological transitions. *Research Policy,* **37**(9), 1436-1445.

Geudens P. J. J. G. (2010). Dutch Drinking Water Statistics 2008. Vewin, Rijswijk, The Netherlands.

Gharajedaghi J. (1999). System Thinking, Managing Chaos and Complexity: A Platform for Designing Business Architecture. Butterworth-Heinemann, Woburn, MA, USA.

Gibbons A. (1993). How the akkadian empire was hung out to dry. *Science,* **261**(5124), 985.

Gibbons M., Limoges C., Nowotny H., Schwartzman S., Scott P. and Trow M. (1994). The New Production of Knowledge: The Dynamics of Science and Research in Contemporary Societies. Sage, London, UK.

Giddens A. (1999). Risk and responsibility. *The Modern Law Review,* **62**(1), 1-10.

Gjesme T. (1983). On the concept of future time orientation: considerations of some functions' and measurements' implications. *International Journal of Psychology,* **18**(4), 443-461.

Goffman E. (1959). The Presentation of Self in Everyday Life. University of Edinburgh Social Sciences Research Centre, Anchor Books edition, Edinburgh, UK.

Goldemberg J. (1998). Leapfrog energy technologies. *Energy Policy,* **26**(10), 729-742.

Gould S. J. (1987). Time's Arrow, Time's Cycle. Myth and Metaphor in the Discovery of Geological Time. Harvard University Press, Cambridge, MA, USA.

Graves T. D. (1967). Psychological acculturation in a tri-ethnic community. *Southwestern Journal of Anthropology,* **23**(4), 337-350.

Gray J. (2013). The Silence of Animals: On Progress and Other Modern Myths. Farrar, Straus and Giroux, New York, USA.

Guba E. G. and Lincoln Y. S. (2005). Paradigmatic controversies, contradictions, and emerging confluences. In: The Sage Publications, Handbook of Qualitative Research, N. K. Denzin and Y. S. Lincoln (eds.), 3rd edn, Thousand Oaks, California, USA, pp. 195-196.

Gunderson L. H. and Holling C. S. (2001). Panarchy: Understanding Transformations in Human and Natural Systems. Island Press, Washington, D.C., USA.

Haith A. M., Reppert T. R. and Shadmehr R. (2012). Evidence for hyperbolic temporal discounting of reward in control of movements. *The Journal of Neuroscience,* **32**(34),11727-11736.

Hambleton R. K., Merenda P. and Spielberger C. (eds.) (2005). Adapting Educational and Psychological Tests for Cross-Cultural Assessment. Lawrence S. Erlbaum Publishers, Hillsdale, MI, USA.

Hancock P. A. (2002). The time of your life. *KronoScope,* **2**(2), 135-165.

Hancock P. A. and Shahnami N. (2010). Memory as a string of pearls. *KronoScope,* **10**(2), 77-82.

Hardisty D. J., Appelt K. C. and Weber E. U. (2013). Good or bad, we want it now: fixed-cost present bias for gains and losses explains magnitude asymmetries in intertemporal choice. *Journal of Behavioral Decision Making,* **26**(4), 348-361.

He X. D. and Zhou X. Y. (2013). Hope, fear, and aspirations. *Mathematical Finance*, doi: 10.1111/mafi.12044. http://people.maths.ox.ac.uk/~zhouxy/download/Hope.pdf

Heffernan J. B. (2008). Wetlands as an alternative stable state in desert streams. *Ecology*, **89**, 1261-1271.

Heideman K., Rostad L. and Youssef M. (eds.) (2012). Has the Arab Spring lived up to expectations? The views of 40 experts from the Middle East, North Africa, Europe, and the United States. *Viewpoints,* **12**, 11-13. Woodrow Wilson International Center for Scholars.

Heine S. J. (2001). Self as cultural product: an examination of East Asian and North American selves. *Journal of Personality,* **69**(6), 881-905.

Heine S. J., Lehman D. R., Peng K. and Greenholtz J. (2002). What's wrong with cross-cultural comparisons of subjective Likert scales? The reference-group effect. *Journal of Personality and Social Psychology,* **82**(6), 903-918.

Hessels L. K. and Van Lente H. (2008). Re-thinking new knowledge production: a literature review and a research agenda. *Research Policy,* **37**(4), 740-760.

Hiwasaki L. (2012). 'Water for life'... Water for whose life? Water, cultural diversity and sustainable development in the United Nations. In: Water, Cultural Diversity, and Global Environmental Change, B. R. Johnston, L. Hiwasaki, I. J. Klaver, A. R. Castillo, and V. Strang (eds.), Springer, The Netherlands, pp. 509-531.

Hoekstra A. Y. and Chapagain A. K. (2007). Water footprints of nations: water use by people as a function of their consumption pattern. *Water Resource Management,* **21**, 35-48.

Hoff H. (2011). Understanding the Nexus, Background Paper for the Bonn 2011 Conference: The Water, Energy and Food Security Nexus, Stockholm Environment Institute, Stockholm.

Hofstede G., Hofstede G. J. and Minkov M. (2010). Cultures and Organisations: Software of the Mind: Intercultural Cooperation and Its Importance for Survival, 3rd edn, McGraw Hill, London, UK.

Hofstede G., Van Deusen C. A., Mueller C. B., Charles T. A. and the Business Goals Network. (2002). What goals do business leaders pursue? A study in fifteen countries. *Journal of International Business Studies,* **33**, 785-803.

Hoogstra M. A. (2008). Coping with the Long Term: An Empirical Analysis of Time Perspectives, Time Orientations, and Temporal Uncertainty in Forestry. PhD thesis, Forest and Nature Conservation Policy Group, Wageningen University, The Netherlands.

Hox J. (2010). Multilevel Analysis: Techniques and Applications, 2nd edn, Routledge, New York, NY, USA.

Hulme M. (2009). Why We Disagree about Climate Change: Understanding Controversy, Inaction and Opportunity. Cambridge University Press, Cambridge, UK.

Husserl E. G. (Brough J. B. Trans.) (1928). On the Phenomenology of the Consciousness of Internal Time.: Kluwer, Dordrecht, The Netherlands.

Husserl E. G. (Cairns D. Trans.) (1929). Formal and Transcendental Logic. Nijhoff, The Hague, The Nethelands.

IPCC (2013). Human Influence on Climate Clear, IPCC Report Says [Press release Intergovernmental Panel on Climate Change]. http://www.ipcc.ch/news_and_events/docs/ar5/press_release_ar5_wgi_en.pdf. 28 September 2013.

Jarymowicz M. and Bar-Tal D. (2006). The dominance of fear over hope in the life of individuals and collectives. *European Journal of Social Psychology,* **36**(3), 367-392.

Jasanoff S. (1994). The Fifth Branch: Science Advisers as Policymakers. Harvard University Press, Cambridge, MA, USA.

Johnson M. (ed.) (1927). More Maxims of Mark, Privately printed.

Jones J. M. (1994). An exploration of temporality in human behavior. In: Beliefs, Reasoning, and Decision Making: Psychologic in Honor of Bob Abelson, R. C. Schank and E. Langer (eds.), Erlbaum, Hillsdale, MI, USA, pp. 389-411.

Jones R. N. and Preston B. L. (2011). Adaptation and risk management. *WIREs Climate Change,* **2**, 296-308.

Kahneman D., Krueger A. B., Schkade D. A., Schwarz N. and Stone A. A. (2004). A survey method for characterizing daily life experience: the day reconstruction method. *Science,* **304**(5702), 1776-1780.

Kashima E. S. and Kashima Y. (1998). Culture and language: the case of cultural dimensions and personal pronoun use. *Journal of Cross-Cultural Psychology,* **29**, 461-486.

Katsumata T. (2007). Application of concept of time perspective to psychotherapy. *Komazawa Annual Reports of Psychology,* **9**, 43-48.

Keough K., Zimbardo P. and Boyd J. (1999). Who's smoking, drinking, and using drugs? Time perspective as a predictor of substance use. *Basic and Applied Social Psychology*, **2**, 149-164.

Ker Rault P. A. and Jeffrey P. (2008). Deconstructing public participation in the Water Framework Directive: implementation and compliance with the letter or with the spirit of the law? *Water and Environment Journal,* **22**, 241-249.

Kinsbourne M. and Hicks R. E. (1990). The extended present: evidence from time estimation by amnesics and normals. In: Neuropsychological Impairments of Short-Term Memory, T. Shallice (ed.), Cambridge University Press, Cambridge, UK, pp. 319–329.

Klineberg S. L. (1967). Changes in outlook on the future between childhood and adolescence. *Journal of Personality and Social Psychology*, **7**, 185-193.

Korthals M. (1994). *Duurzaamheid en Democratie* (Sustainability and democracy). Boom, Amsterdam, The Netherlands.

Lach D., Rayner S. and Ingram H. (2005). Taming the waters: strategies to domesticate the wicked problems of water resource management. *International Journal of Water*, **3**(1), 1-17.

Lachman M. E., Rocke C., Rosnick C. and Ryff C. D. (2008). Realism and illusion in Americans' temporal views of their life satisfaction: age differences in trajectories of life satisfaction. *Psychological Science*, **19**(9), 889-897.

Lang D. J., Wiek A., Bergmann M., Stauffacher M., Martens P., Moll P., Swilling M. and Thomas C. J. (2012). Transdisciplinary research in sustainability science: practice, principles, and challenges. *Sustainability Science*, **7**(1), 25-43.

Lee N. and Brown S. (1994). Otherness and the actor network. *American Behavioral Scientist*, **37**(6), 772-790.

Lessing E. E. (1968). Demographic, developmental, and personality correlates of length of future time perspective. *Journal of Personality*, **36**, 193-201.

Lessing E. E. (1972). Extension of personal future time perspective, age, and life satisfaction of children and adolescents. *Developmental Psychology*, **6**(3), 457-468.

Levine J. H. (1972). The sphere of influence. *American Sociological Review*, **37**(1), 14-27.

Levine R. V. (1997). *A Geography of Time*. Basic Books, New York, NY, USA.

Lichtenstein R., Alexander J. A., Jinnett K. and Ullman E. (1997). Embedded intergroup relations in interdisciplinary teams: effects on perceptions of level of team integration. *Journal of Applied Behavioral Science*, **33**(4), 413-434.

Loarie S. R., Duffy P. B., Hamilton H., Asner G. P., Field C. B. and Ackerly D. D. (2009). The velocity of Climate Change. *Nature*, **462**(7276), 1052-1055.

Lomberg B. (ed.) (2004). Global Crisis, Global Solutions. Cambridge University Press, Cambridge, UK.

Lopes L. L. (1987). Between hope and fear: the psychology of risk. In: Advances in Experimental Social Psychology, L. Berkowitz (ed.), Academic Press, San Diego, CA, USA.

Maguire S. and McKelvey B. (1999). Complexity and management: moving from fad to firm foundations. *Emergence*, **1**(2), 19-61

Masini E. B. (2001). New challenges for futures studies. *Futures*, **33**(7), 637-647.

Mazur J. E. (1987). An adjusting procedure for studying delayed reinforcement. In: Quantitative Analyses of Behavior: Vol. 5. The Effect of Delay and of Intervening Events on Reinforcement Value, M. L. Commons, J. E. Mazur, J. A. Nevin and H. Rachlin (eds.), Erlbaum, Hillsdale, NJ, USA, pp. 55–73.

Mckay M. T., Ballantyne N., Goudie A. J., Sumnall H. R. and Cole J. C. (2012). Here for a good time, not a long time: decision making, future consequences and alcohol use among Northern Irish adolescents. *Journal of Substance Use*, **17**, 1-18.

Meadows D. H., Meadows D. L., Randers J. and Behrens III W. W. (1972). The Limits to Growth. Universe Books, New York, NY, USA.

Medema W., McIntosh B. S. and Jeffrey P. J. (2008). From premise to practice: a critical assessment of integrated water resources management and adaptive management approaches in the water sector. *Ecology and Society*, **13**(2), 29.

Meijer M. H. (2007). Intertwining Uncertainty Analysis and Decision-making about Drinking Water Infrastructure. Unpublished PhD thesis, Delft University of Technology, Delft, The Netherlands.

Minkov M. (2012). Cross-Cultural Analysis: The Science and Art of Comparing the World's Modern Societies and Their Cultures. Sage, London, UK.

Mintzberg H. (1994). The Rise and Fall of Strategic Planning. The Free Press, New York, NY, USA.

Mitroff I. A. and Turoff M. (1975). Philosophical and methodological foundations of delphi. In: The Delphi Method Techniques and Applications, H. A. Linstone and M. Turoff (eds.), Addison-Wesley, Michigan, IN, USA.

Molle F. (2008). Nirvana concepts, narratives and policy models: insights from the water sector. *Water Alternatives*, **1**(1), 131-156.

Monks F. J. (1967). Time perspective and a psychological variable. *Archiv für die Gesamte Psychologie*, **119**(1-2), 131-161.

Moss R. H., Edmonds J. A., Hibbard K. A., Manning M. R., Rose S. K., van Vuuren D. P. and Wilbanks T. J. (2010). The next generation of scenarios for climate change research and assessment. *Nature,* **463**(7282), 747-756.

Mosse D. (2004). Is good policy unimplementable? Reflections on the ethnography of aid policy and practice. *Development and Change,* **35**(4), 639-671.

Mosse D. and Sivan M. (2003). The Rule of Water: Statecraft, Ecology and Collective Action in South India. Oxford University Press, Oxford, UK.

Muro M. and Jeffrey P. (2008). A critical review of the theory and application of social learning in participatory natural resource management processes. *Journal of Environmental Planning and Management,* **51**(3), 325-344.

Muro M. and Jeffrey P. (2012). Time to talk? How the structure of dialog processes shapes stakeholder learning in participatory water resources management. *Ecology and Society,* **17**(1), 3.

Nachtnebel H. P. (2002). Irreversibility and sustainability in water resources systems. In: *Risk, Reliability, Uncertainty, and Robustness of Water Resource Systems,* J. J. Bogardi and Z. W. Kundzewicz (eds.), Cambridge University Press, Cambridge, UK, pp. 181-189.

Nanni G. (2011). Time, empire and resistance in settler-colonial Victoria. *Time and Society,* **20**(1), 5-33.

Newell A. (1980). Reasoning, problem solving and decision processes: the problem space as a fundamental category. In: Attention and Performance VIII, R. S. Nickerson (ed.), Lawrence Erlbaum, Hillsdale, MI, USA.

Nisbet M. C. (2010). Framing science: a new paradigm in public engagement. In: New Agendas in Science Communication, L. A. Kahlor and P. Stout (eds.), Taylor and Francis Publishers, New York, NY, USA, pp. 40-66.

Nisbet R. (1979). The idea of progress: a bibliographical essay. *Literature of Liberty: A Review of Contemporary Liberal Thought,* **2**(1), 13.

Norenzayan A., Choi I. and Peng K. (2010). Perception and cognition. In: Handbook of Cultural Psychology, S. Kitayama and D. Cohen (eds.), Guilford Press, New York, NY, USA, pp. 569-594.

Norenzayan A. and Schwarz N. (1999). Telling what they want to know: participants tailor causal attributions to researchers' interests. *European Journal of Social Psychology,* **29**, 1011-1020.

Noyes R. (1980). The time horizon of planned social change: why Utopian movements always promise amelioration in the future. *American Journal of Economics and Sociology,* **39**(1), 65-77.

Nuttin J. R. (1964). The future time perspective in human motivation and learning. *Acta Psychologies,* **23**, 60-82.

Nuttin J. (1984). Motivation, Planning, and Action: A Relational Theory of Behavioral Dynamics. Erlbaum, Hillsdale, MI, USA.

Nuttin J. and Lens W. (eds.) (1985). Future Time Perspective and Motivation: Theory and Research Method. Leuven University Press, Leuven, Belgium.

Odion D. C., Moritz M. A. and DellaSala D. A. (2009). Alternative community states maintained by fire in the Klamath Mountains, USA. *Journal of Ecology,* **98**, 96-105.

Okhuysen G. A., Galinsky A. D. and Uptigrove T. A. (2003). Saving the worst for last: the effect of time horizon on the efficiency of negotiating benefits and burdens. *Organizational Behavior and Human Decision Processes,* **91**, 269-279.

Oyserman D., Coon H. M. and Kemmelmeier M. (2002). Rethinking individualism and collectivism: evaluation of theoretical assumptions and meta-analysis. *Psychological Bulletin,* **128**, 3-72.

Pahl-Wostl C., Jeffrey P. and Sendzimir J. (2011). Adaptive and integrated management of water resources. In: Water Resources Planning and Management, R. Q. Grafton and K. Hussey (eds.),Cambridge University Press, Cambridge, UK, pp. 292-309.

Pahl-Wostl C., Kabat P. and Möltgen J. (eds.) (2007a). Adaptive and Integrated Water Management: Coping with Complexity and Uncertainty. Springer, Berlin, Germany.

Pahl-Wostl C., Sendzimir J., Jeffrey P., Aerts J., Berkamp G. and Cross K. (2007b). Managing change toward adaptive water management through social learning. *Ecology and Society,* **12**(2), 30-38.

Pahl-Wostl C., Tàbara D., Bouwen R., Craps M., Dewulf A., Mostert E., Ridder D. and Taillieu T. (2008). The importance of social learning and culture for sustainable water management. *Ecological Economics,* **64**, 484-495.

Patomäki H. (2006). Realist ontology for futures studies. *Journal of Critical Realism,* **5**(1), 1-31.

Pearson L. J., Coggan A., Proctor W. and Smith T. F. (2010). A sustainable decision support framework for urban water management. *Water Resources Management,* **24**(2), 363-376.

Peng K., Nisbett R. E. and Wong N. Y. C. (1997). Validity problems comparing values across cultures and possible solutions. *Psychological Methods,* **2**, 329-344.

Perkins R. (2003). Environmental leapfrogging in developing countries: a critical assessment and reconstruction. *Natural Resources Forum,* **27**(3), 177-188.

Piaget J. (Pomerans A. J. Trans.) (1969). The Child's Conception of Time. Ballantine Books, New York, NY, USA.

Pielke Jr R. A. (2007). The Honest Broker: Making Sense of Science in Policy and Politics. Cambridge University Press, Cambridge, UK.

Pinchbeck D. (2007). 2012: *The Return of Quetzalcoatl.* Penguin.

Pink B. (2012). 2012 Year Book Australia Number 92. Australian Bureau of Statistics, Canberra, Australia. ABS Catalogue No. 1301.0.

Polak F. L. (1973). The Image of the Future. Elsevier Scientific Publishing Company, Amsterdam, The Netherlands.

Pregernig M. (2002). Perceptions, not facts: how forestry professionals decide on the restoration of degraded forest ecosystems. *Journal of Environmental Planning and Management*, **45**(1), 25-38.

Raihani N. and Aitken D. (2011). Uncertainty, rationality and cooperation in the context of climate change. *Climatic Change,* **108**, 47-55.

Raju C. K. (2003). The Eleven Pictures of Time: The Physics, Philosophy, and Politics of Time Beliefs. Sage, London, UK.

Rappaport A. (2005). The economics of short-term performance obsession. *Financial Analysts Journal,* **61**(3), 65-79.

Rappaport H., Enrich K. and Wilson A. (1985). Relation between ego identity and temporal perspective. *Journal of Personality and Social Psychology*, **48**(6), 1609-1620.

Raudenbush S. W. and Liu X. (2000). Statistical power and optimal design for multisite randomized trials. *Psychological Methods,* **5**, 199-213.

Ravetz J. (2004). The post-normal science of precaution. *Futures,* **36**, 347-357.

Read D., Bostrom A., Morgan M. G., Fischhoff B. and Smuts T. (1994). What do people know about global climate change? 2. Survey studies of educated laypeople. *Risk Analysis,* **14**(6), 971-982.

Rip A. (2002). Regional innovation systems and the advent of strategic science. *Journal of Technology Transfer,* **27**, 123-131.

Rittel H. W. J. and Webber M. M. (1973). Dilemmas in a general theory of planning. *Policy Sciences,* **4**, 155-169.

Roccas S. and Brewer M. B. (2002). Social identity complexity. *Personality and Social Psychology Review,* **6**(2), 88-106.

Rockenbach B. and Milinski M. (2006). The efficient interaction of indirect reciprocity and costly punishment. *Nature,* **444**, 718-723.

Rockström J., Steffen W., Noone K., Persson A., Chapin F. S., Lambin E. F., Lenton T. M., Scheffer M., Folke C., Schellnhuber H. J., Nykvist B., De Wit C. A., Hughes T., Van der Leeuw S., Rodhe H., Sörlin S., Snyder P. K., Costanza R., Svedin U., Falkenmark M., Karlberg L., Corell R. W., Fabry V. J., Hansen J., Walker B., Liverman D., Richardson K., Crutzen P. and Foley J. A. (2009). A safe operating space for humanity. *Nature,* **461**, 472-475.

Rowe A. and Hogarth A. (2005). Use of complex adaptive systems metaphor to achieve professional and organizational change. *Journal of Advanced Nursing*, **51**(4), 396-405.

Sanna L. J. and Schwarz N. (2003). Debiasing the hindsight bias: the role of accessibility experiences and (mis) attributions. *Journal of Experimental Social Psychology*, **39**, 287-295.

Sarewitza D. and Pielke Jr R. (1999). Prediction in science and policy. *Technology in Society,* **21**, 121-133.

Scheffer M. (2009). Critical Transitions in Nature and Society. Princeton University Press, Princeton, NJ.

Scheffer M., Bascompte J., Brock W. A., Brovkin V., Carpenter S. R., Dakos V., Held H., Van Nes E. H., Rietkerk M. and Sugihara G. (2009). Early-warning signals for critical transitions. *Nature,* **461**(7260), 53-59.

Scheffer M. and Carpenter S. R. (2003). Catastrophic regime shifts in ecosystems: linking theory to observation. *Trends in Ecology and Evolution,* **18**(12), 648-656.

Scheffer M., Hosper S. H., Meijer M. L., Moss B. and Jeppesen E. (1993). Alternative equilibria in shallow lakes. *Trends in Ecology and Evolution*, **8**, 275-279.

Scheuerman W. E. (2004). Liberal Democracy and the Social Acceleration of Time. The Johns Hopkins University Press, Baltimore, MD, USA.

Schlenker B. R. (1985). Identity and self-identification. In: The Self and Social Life, B. R. Schlenker (ed.), McGraw-Hill, New York, NY, USA, pp. 65-99.

Seale C. (1999). The Quality of Qualitative Research. Blackwell, Oxford, UK.

Segrave A. J., Büscher C. and Frijns J. (2011). Participatory futures research and social learning for integrated watershed management. In: Environmental Security in Watersheds: The Sea of Azov. NATO Science for Peace and Security Series C: Environmental Security, V. Lagutov (ed.), Springer, Dordrecht, The Netherlands, pp. 241-253.

Segrave A. J., Van der Zouwen M. W. and Van Vierssen W. (2013a). Water planning: from what time perspective? *Technological Forecasting and Social Change,* http://dx.doi.org/10.1016/j.techfore.2013.08.019.

Segrave A. J., Wen B., Van der Zouwen M. W. and Van Vierssen W. (2013b). Who are the aliens actually? Mapping problem perception in the water sector. [Submitted for review.]

Shore B. (1996). Culture in Mind: Cognition, Culture, and the Problem of Meaning. Oxford University Press, Oxford, UK.

Simons J., Vansteenkiste M., Lens W. and Lacante M. (2004). Placing motivation and future time perspective theory in a temporal perspective. *Educational Psychology Review,* **16**(2), 121-139.

Simpleton J. (2009). Nationalization of Government, Speech presented at the Preservation of Democracy Society, Knoxville, TN, USA.

Smith E. R. (1999). Affective and cognitive implications of a group becoming part of the self: new models of prejudice and of the self concept. In: Social Identity and Social Cognition, D. Abrams and M. A. Hogg (eds.), Blackwell, Oxford, UK, pp. 183-196.

Sterling S. (2007). Riding the storm: towards a connective cultural consciousness. In: Social Learning: Towards a Sustainable World, A. E. J. Wals (ed.), Wageningen Academic Publishers, Wageningen, The Netherlands, p. 63.

Stern N. (2006). What is the economics of climate change? *World Economics,* **7**(2), 1-10.

Storey K. (2001). Fly-in/fly-out and fly-over: mining and regional development in Western Australia. *Australian Geographer,* **32**(2), 133-148.

Strang V. (2004). The Meaning of Water. Berg, Oxford, UK.

Strathman A., Gleicher F., Boninger D. S. and Edwards C. S. (1994). The consideration of future consequences: weighing immediate and distant outcomes of behaviour. *Journal of Personality and Social Psychology,* **66**(4), 742-752.

Streffer C., Gethmann C. F., Kamp G., Kröger W., Rehbinder E., Renn O. and Röhlig K. J. (2012). Radioactive Waste: Technical and Normative Aspects of Its Disposal. Springer, Heidelberg, Germany.

Struble Jr R. (1979). House turnover and the principle of rotation. *Political Science Quarterly,* **94**, 649-667.

Stump R. L., Gong W. and Li Z. (2008). Exploring the digital divide in mobile-phone adoption levels across countries do population socioeconomic traits operate in the same manner as their individual-level demographic counterparts? *Journal of Macromarketing,* **28**(4), 397-412.

Stuyfzand P. J. (1995). The impact of land reclamation on groundwater quality and future drinking water supply in the Netherlands. *Water Science and Technology*, **31**(8), 47-57.

Suarez F. F. and Oliva R. (2005). Environmental change and organizational transformation. *Industrial and Corporate Change,* **14**(6), 1017-1041.

Suding K. N., Gross K. L. and Houseman G. R. (2004). Alternative states and positive feedbacks in restoration ecology. *Trends in Ecology and Evolution*, **19**, 46-53.

Sultana F. (2013). Water, technology, and development: transformations of development technonatures in changing waterscapes. *Environment and Planning D: Society and Space*, **31**(2), 337-353.

Sutherland W. J. and Woodroof H. J. (2009). The need for environmental horizon scanning. *Trends in Ecology and Evolution,* **24**(10), 523-527.

Swann Jr. W. B. (1987). Identity negotiation: where two roads meet. *Journal of Personality and Social Psychology,* **53**, 1038-1051.

Swann Jr. W. B., Johnson R. E. and Bosson J. K. Identity negotiation at work. *Research in organizational behavior,* **29** (2009): 81-109.

Tainter J. (1990). The Collapse of Complex Societies. Cambridge University Press, Cambridge, UK.

Thaler R. (1980). Toward a positive theory of consumer choice. *Journal of Economic Behavior and Organization,* **1**, 39-60.

Thompson-Klein J., Grossenbacher-Mansuy W. and Häberli R. (2004). Transdisciplinarity: Joint Problem Solving among Science, Technology, and Society: An Effective Way for Managing Complexity. Birkhäuser, Boston, MA, USA.

Tickner J. A. (ed.) (2003). Precaution, Environmental Science, and Preventive Public Policy. Island Press, London, UK.

Ting-Toomey S. (1993). Communication resourcefulness: an identity-negotiation perspective. In: Intercultural Communication Competence, R. Wiseman and J. Koester (eds.), Sage, Newbury Park, CA, USA, pp. 72-111.

Triandis H. C. (1996). The psychological measurement of cultural syndromes. *American Psychologist,* **51**(4), 407-415.

Tversky A. and Kahneman D. (1981). The framing of decisions and the psychology of choice. *Science,* **211**(4481), 453-458.

Tukker A. (2005). Leapfrogging into the future: developing for sustainability. *International Journal of Innovation and Sustainable Development,* **1**(1), 65-84.

Van Andel J. and Aronson J. (eds.) (2012). Restoration Ecology: The New Frontier. Wiley-Interscience.

Van Asselt M. B. A., Faas A., Van der Molen F. and Veenman S. A. (eds.) (2010a). *Uit Zicht: Toekomstverkennen met Beleid* (From Visibility: Explore Future of Policy). Amsterdam University Press, Amsterdam, The Netherlands.

Van Asselt M. B. A., Van 't Klooster S. A., Van Notten P. W. F. and Smits L. A. (2010b). Foresight in Action: Developing Policy-Oriented Scenarios. Earthscan, London, UK.

Van der Brugge R., Rotmans J. and Loorbach D. (2005). The transition in Dutch water management. *Regional Environmental Change,* **5**, 164-176.

Van 't Klooster S. A. (2007). *Toekomstverkenning: ambities en de praktijk. Een etnografische studie naar de productie van toekomstkennis bij het Ruimtelijk Planbureau (RPB)* (Futures research: ambitions and practices: an ethnographical study into the production of knowledge about the future by the bureau of spatial planning). Eburon, Delft, The Netherlands.

Van Vierssen W. (2012). *Water@work! Over een kennissysteem met kleine spelers en grote ambities* (Water@work! A knowledge system with small players and big ambitions), Inaugural address to mark becoming professor in the field Science System Assessment of the water-related research at Delft University of Technology, The Netherlands.

Wackernagel M. and Rees W. (1996). Our Ecological Footprint: Reducing Human Impact on the Earth. New Society Publishers, Gabriola Island, BC, Canada.

Wageman R., Gardner H. and Mortensen M. (2012a). Teams have changed: catching up to the future. *Industrial and Organizational Psychology,* **5**(1), 48-52.

Wageman R., Gardner H. and Mortensen M. (2012b). The changing ecology of teams: new directions for teams research. *Journal of Organizational Behavior,* **33**, 301-315.

Wallace M. (1956). Future time perspective in schizophrenia. *Journal of Abnormal Psychology,* **52**, 240-245.

Walton J. S. (2008). Scanning beyond the horizon: exploring the ontological and epistemological basis for scenario planning. *Advances in Developing Human Resources,* **10**(2), 147-165.

Waughray D. (ed.) (2011). Water Security: The Water Food Energy Climate Nexus. World Economic Forum Initiative report. Island Press, Washington, D.C., USA.

Webler T. and Renn O. (1995). A brief primer on participation: philosophy and practice. In: Fairness and Competence in Citizen Participation: Evaluating Models for Environmental Discourse, O. Renn, T. Webler, and P. Wiedemann (eds.), Kluwer, Dordrecht, The Netherlands, p. 17.

WFD 2000/60/EC (2000). Directive of the European Parliament and of the Council establishing a framework for the community action in the field of water policy. *Official Journal* (OJ L 327), 1-73.

Whitney J. C. and Smith R. A. (1983). Effects of group cohesiveness on attitude polarization and the acquisition of knowledge in a strategic planning context. *Journal of Marketing Research,* **20**(2), 167-176.

Wilby R. L. and Wigley T. M. L. (1997). Downscaling general circulation model output: a review of methods and limitations. *Progress in Physical Geography,* **21**(4), 530-548.

Wohlford P. and Herrera J. A. (1970). TAT stimulus cues and extension of personal time. *Journal of Projective Techniques and Personality Assessment,* **34**, 31-37.
Wood C. H. (2008). Time, cycles and tempos in social-ecological research and environmental policy. *Time and Society,* **17**(2), 261-282.
Wood S. (2011). Four varieties of social responsibility: making sense of the 'sphere of influence' and 'leverage' debate via the case of ISO 26000. *Osgoode CLPE Research Paper Series*, **7**(4), 14-35.
Woolley A. W., Chabris C. F., Pentland A., Hashmi N. and Malone T. W. (2010). Evidence for a collective intelligence factor in the performance of human groups. *Science,* **330**(6004), 686-688.
Wright R. (2004). A Short History of Progress. House of Anansi Press, Toronto, Canada.
Ylijoki O. H. and Mäntylä H. (2003). Conflicting time perspectives in academic work. *Time and Society,* **12**(1), 55-78.
Young O. R., Berkhout F., Gallopin G. C., Janssen M. A., Ostrom E. and Van der Leeuw S. (2006). The globalization of socio-ecological systems: an agenda for scientific research. *Global Environmental Change,* **16**(3), 304.
Youssef C. M. and Luthans F. (2007). Positive organizational behavior in the workplace: the impact of hope, optimism, and resilience. *Journal of Management,* **33**(5), 774-800.
Zimbardo P. and Boyd J. (2008). The Time Paradox: The New Psychology of Time That Will Change Your Life. Simon and Schuster, New York, NY, USA.

Lightning Source UK Ltd.
Milton Keynes UK
UKOW06f1834041214

242668UK00003B/212/P